I0516066

Tucholsky Wagner Zola Scott Sydow Freud Schlegel
Turgenev Wallace Fonatne
Twain Walther von der Vogelweide Fouqué Friedrich II. von Preußen
Weber Freiligrath Frey
Fechner Fichte Weiße Rose von Fallersleben Kant Ernst Frommel
Hölderlin Richthofen
Engels Fielding Eichendorff Tacitus Dumas
Fehrs Faber Flaubert
Maximilian I. von Habsburg Fock Eliasberg Zweig Ebner Eschenbach
Feuerbach Ewald Eliot Vergil
Goethe Elisabeth von Österreich London
Mendelssohn Balzac Shakespeare Dostojewski Ganghofer
Trackl Stevenson Lichtenberg Rathenau Doyle Gjellerup
Mommsen Tolstoi Hambruch
Thoma Lenz Hanrieder Droste-Hülshoff
Dach Verne von Arnim Hägele Hauff Humboldt
Karrillon Reuter Rousseau Hagen Hauptmann Gautier
Garschin Defoe Hebbel Baudelaire
Damaschke Descartes
Hegel Kussmaul Herder
Wolfram von Eschenbach Darwin Dickens Schopenhauer Rilke George
Bronner Melville Grimm Jerome Bebel Proust
Campe Horváth Aristoteles Voltaire Federer
Bismarck Vigny Barlach Heine Herodot
Gengenbach Grillparzer
Storm Casanova Lessing Tersteegen Gilm Georgy
Chamberlain Langbein Gryphius
Brentano Claudius Schiller Lafontaine
Strachwitz Kralik Iffland Sokrates
Katharina II. von Rußland Bellamy Schilling
Gerstäcker Raabe Gibbon Tschechow
Löns Hesse Hoffmann Gogol Wilde Gleim Vulpius
Luther Heym Hofmannsthal Klee Hölty Morgenstern Goedicke
Roth Heyse Klopstock Kleist
Luxemburg Puschkin Homer Mörike
Machiavelli La Roche Horaz Musil
Navarra Aurel Musset Kierkegaard Kraft Kraus
Lamprecht Kind Kirchhoff Hugo Moltke
Nestroy Marie de France
Nietzsche Nansen Laotse Ipsen Liebknecht
Marx Lassalle Gorki Klett Leibniz Ringelnatz
von Ossietzky May vom Stein Lawrence Irving
Petalozzi Knigge
Platon Pückler Michelangelo Kafka
Sachs Poe Liebermann Kock
Korolenko
de Sade Praetorius Mistral Zetkin

The publishing house tredition has created the series **TREDITION CLASSICS**. It contains classical literature works from over two thousand years. Most of these titles have been out of print and off the bookstore shelves for decades.

The book series is intended to preserve the cultural legacy and to promote the timeless works of classical literature. As a reader of a **TREDITION CLASSICS** book, the reader supports the mission to save many of the amazing works of world literature from oblivion.

The symbol of **TREDITION CLASSICS** is Johannes Gutenberg (1400 – 1468), the inventor of movable type printing.

With the series, tradition intends to make thousands of international literature classics available in printed format again – worldwide.

All books are available at book retailers worldwide in paperback and in hardcover. For more information please visit: www.tredition.com

tredition was established in 2006 by Sandra Latusseck and Soenke Schulz. Based in Hamburg, Germany, tredition offers publishing solutions to authors and publishing houses, combined with worldwide distribution of printed and digital book content. tredition is uniquely positioned to enable authors and publishing houses to create books on their own terms and without conventional manufacturing risks.

For more information please visit: www.tredition.com

Home Taxidermy for Pleasure and Profit A Guide for Those Who Wish to Prepare and Mount Animals, Birds, Fish, Reptiles, etc., for Home, Den, or Office Decoration

Albert B. Farnham

Imprint

This book is part of the TREDITION CLASSICS series.

Author: Albert B. Farnham
Cover design: toepferschumann, Berlin (Germany)

Publisher: tredition GmbH, Hamburg (Germany)
ISBN: 978-3-8491-8976-1

www.tredition.com
www.tredition.de

Copyright:
The content of this book is sourced from the public domain.

The intention of the TREDITION CLASSICS series is to make world literature in the public domain available in printed format. Literary enthusiasts and organizations worldwide have scanned and digitally edited the original texts. tredition has subsequently formatted and redesigned the content into a modern reading layout. Therefore, we cannot guarantee the exact reproduction of the original format of a particular historic edition. Please also note that no modifications have been made to the spelling, therefore it may differ from the orthography used today.

"WHOO" SAID WISE OLD OWL.

Home Taxidermy for Pleasure and Profit

A Guide for those who wish to prepare
and mount animals, birds, fish,
reptiles, etc., for home, den,
or office decoration

By
ALBERT B. FARNHAM, Taxidermist

ALBERT B. FARNHAM, Author

CONTENTS

Chapter.

	Introductory
I.	History of the Art
II.	Outfit — Tools and Materials
III.	Preservative Preparations, Formula, etc.
IV.	Panels, Shields and Natural and Artificial Mounts
V.	Field Work, Collecting
VI.	Skinning and Preserving Skins
VII.	Making Scientific Skins
VIII.	Preparing Dry and Wet Skins for Mounting
IX.	Mounting Small and Medium Birds
X.	Mounting Large Birds
XI.	Tanning, Cleaning, and Poisoning Skins
XII.	Making Animal Fur Rugs
XIII.	Fur Robes and How to Make Them
XIV.	Mounting Entire Small Animals
XV.	Mounting Large Animals
XVI.	Mounting Heads of Small Animals, Birds and Fish
XVII.	Mounting Heads of Large Game
XVIII.	Mounting Horns and Antlers
XIX.	Mounting Feet and Hoofs
XX.	Mounting Fish
XXI.	Mounting Fish — Baumgartel Method
XXII.	Mounting Reptiles, Frogs and Toads
XXIII.	Skulls and Skeletons
XXIV.	Sportsmen's Trophies
XXV.	Odds and Ends, Taxidermy Novelties

XXVI.	GROUPS AND GROUPING
XXVII.	ANIMAL ANATOMY
XXVIII.	CASTING AND MODELLING
XXIX.	MARKET TROPHY HUNTING
XXX.	COLLECTING AND MOUNTING FOR SALE
XXXI.	PRICES FOR WORK

List of Illustrations

"Whoo? Said Wise Old Owl"
A Specimen of the Early Day
Work Table, Supplies, Tools, Etc.
Home Made Tools
Taxidermists Tools
Taxidermists Tools—Scalpels, Scissors, Stuffers
Egg Drill and Other Tools
Glass Eyes for Animals, Birds, Fish
Sizes of Glass Eyes
Sizes of Glass Eyes (Style 1)
Shields—Various Kinds and Sizes
Shields, Foot and Hall Rack
Gun Rack, Fish and Game Panels, Hall Rack
Some Shields and Panels
Home Made Shield
Small Bird
Marbles Shot Gun and Rifle Combined
The "Stop Thief" or Choke Trap
Outline of Muskrat
Skinning Small Animal for Mounting
Skinning Large Animal for Mounting
Skinning Bird—Breast Cut
Hooded Merganser
Opening Cut on Bird
Scientific Skins, Small Animals and Birds
Foot Skinned Out
Clinching Leg Wires in Artificial Bird Body
Wire Loop for Bird Body

Wiring Leg of Bird
Bird Wired
Bird Wound With Thread
Pose or Position of Certain Birds
Natural Standing Position of Small Bird
Fluttering Position of Small Birds
A Bird of Prey—Limb Position
Spreading Tail of Large Birds
Eagle—Wings Braced up to Dry
Fleshing Knife
Bench Beam
Adjustable Beam
Paper Head Forms—Fox
Paper Head Form—Dog Closed Mouth
Foundation for Tow and Excelsior
Sewing up Bullet Hole
Pinked Lining, Briar Stitched
Sewing Trimming on Rug
Coyote Rug, Open Mouth
Coon Skin Marked to Cut for Robe
Strong Hide (Cattle) Laprobe
Eight Skin Coyote Laprobe
Patagonian Robe of Guanaco Skins
Muskrat Legs Wrapped Ready to Clay
Wiring for Small Animals
Opossum Mounted in Walking Position
Cat Sitting and Watching
Frame for Bear Manikin
Bolting Leg Rods to Stand
Fastening Rods to Back Board
Wild Cat Head Mounted on Shield

Fox Head on Neck Standard
Leopard Head, Artificial
Hawk Head
Sheep Head
Skinning Horned Heads
Horned Heads — Antelope, Deer
Deer Skull on Standard
Neck Board
Paper Head and Ear Forms
Finished Head — Author's Work
Elk and Deer Head Paper Forms
Bolting Shed Antlers to Block
Shed Elk Antlers to be Mounted
A Good Shield Pattern
Deer Antlers, Elk Feet, Bison Horns
Wooden Crook for Deer Foot
Skinned Deer Foot
Deer Foot Ink Well
Deer Foot Thermometer
Deer Foot Hat Rack
Moose Foot Jewel Case
Plaster Mould of Fish — Upper Half
Medallion Fish
Grayling — Results of First Fish Mounting
Eye of Lake Trout
Fish in Mould
Fish in Mould — End View
Fish in Mould — Side View
Fish — Incisions to be Made
Fish — Inside Board
Fish — Sewing up the Skin

Home Made Panel for Fish
Fish Head Mounted — Side View
Fish Head Mounted — Front View
Wiring System for Frog
Skulls — Wolf, Lynx, Otter, Mink
Flying Duck
Timber Wolf Rug, Full Head
Deer Head Hall Rack
Foot Muff Trimmed
Monkey Card Receiver
Squirrels — Grey, Red, Flying, Ground
Water Fowl Head
Three Piece Mould of Head
Making Mould for Half Head
Deer Foot Ink Well and Pen Rack
Mountain Lion or Puma Hide
Spring Lamb? Coon Head
Book Case Ornaments — Crow, Alligator, Owl

INTRODUCTION

This volume of the Pleasure and Profit Library is offered to the hunter, trapper, fisher, vacationist and out of doors people in general. In the study and practice of taxidermy for several years I have failed to find any work written primarily for these every day nature lovers, though they probably handle a greater number of interesting specimens of animal life than all other classes of people.

In view of this fact the following directions and suggestions for preserving various animal forms as objects of use and ornament have been prepared. As a treatise for the scientist or museum preparator it is not intended, there are many books on the art expressly for them, but we hope it may fill a place of its own, acting as a not too dry and technical introduction to the art preservative for those who find life all too short for the many things which are to be done.

Thoroughness, patience, and some love for nature, are qualities highly desirable in this art. Work prepared by one possessing these qualities need not be ashamed and practice will bring skill and perfection of technic.

As a handicraft in which the workman has not been displaced or made secondary by a machine taxidermy is noticeable also, and for many reasons is worthy of its corner in the home work-shop.

In this work also the ladies can take a very effective hand, and numbers have done so; for there is no doubt that a woman's taste and lightness of touch enables her in some branches of taxidermy to far exceed the average man. Especially in the manipulation of frail skins and delicate feathers, in bird taxidermy, is this so.

I have endeavored to give preference to short cuts and time-saving methods where possible in the following matter, and especially hints on saving interesting and valuable specimens temporarily until sufficient leisure is had to do justice to their further preservation. In this connection I have given prominence to the liquid preservative for entire specimens and the methods for preserving skins of birds and animals in a damp and relaxed state ready for mounting at the operator's pleasure.

I would urge the beginner especially, to mount all his specimens as far as possible. Dry scientific skins have their value, perhaps, to the museum or closet naturalist whose chief delight is in multiplying species, but a well mounted skin is a pleasure to all who may see it. Making it a rule to utilize thus all specimens which come to hand would also deter much thoughtless killing in the ranks of the country's already depleted wild life.

Make this a rule and you will avoid friction and show yourself truly a conservationist with the best. In a number of states there are legal restrictions in the way of a license tax imposed on the professional taxidermist. Detailed information of these are found in Game, Fur and Fish Laws of the various states and Canadian provinces. Fur and game animals and birds killed legally during open season may be preserved by the taker for private possession without hindrance anywhere, I think. More explicit details may be had on application to your state fish and game commissioner or warden.

Albert B. Farnham.

Home Taxidermy for Pleasure and Profit

CHAPTER I.

HISTORY.

It is very evident that this art—Taxidermy, preservation or care of skins—had its origin far back before the dawn of written history. There existed then as now the desire to preserve the trophy of the hunter's prowess and skill and the unusual in natural objects.

As far back as five centuries B. C. in the record of the African explorations of Hanno the Carthaginian, an account is given of the discovery of what was evidently the gorilla and the subsequent preservation of their skins, which were, on the return of the voyagers, hung in the temple of Astarte, where they remained until the taking of Carthage in the year 146 B. C.

This, of course, was not the art as we know it now, but shows the beginnings of what might be called the museum idea. The art of embalming as practiced by the ancient Egyptians was, however, effective, not for the purpose of having the specimens look natural, or for exhibition, but to satisfy the superstition of the times, and though a preservative art, hardly to be classed with taxidermy.

In the tombs of that period are found besides the mummies of human beings, countless others of dogs, cats, monkeys, birds, sheep and oxen. There have been a number of efforts made to substitute some form of embalming for present day taxidermy but without much success, for though the body of the specimen may be preserved from decay without removing it from the skin, the subsequent shrinkage and distortion spoil any effect which may have been achieved.

The first attempt at stuffing and mounting birds was said to have been made in Amsterdam in the beginning of the 16th century. The oldest museum specimen in existence, as far as I know, is a rhinoceros in the Royal Museum of Vertebrates in Florence, Italy, said to have been originally mounted in the 16th century.

Probably on account of the necessary knowledge of preservative chemicals, the art seems to have been in the hands of chemists and astrologers, chiefly, during the middle ages, and stuffed animals such as bats, crocodiles, frogs, snakes, lizards, owls, etc., figure in literary descriptions of their abodes. Then as now also, the dining halls of the nobles and wealthy were decorated with heads and horns procured in the hunt.

AN EARLY DAY SPECIMEN.

The first publications on the art seem to have been made in France, in which country and Germany, many still used methods and formulas originated. Though the first volume of instruction in taxidermy was published in the United States as late as 1865, it has been left for the study and ingenuity of American taxidermists to accomplish what is probably work of as high a standard as any in the world.

The Ward establishment at Rochester has turned out many well trained taxidermists, the large museums of the United States are filled with some of the best work of the kind in existence, besides many persons who have engaged in it for commercial purposes or to gratify private tastes. Many of these have made public their methods and modes in various publications. Among these are the works of Batty, Hornaday, Shofeldt, Davie, Rowley, Maynard, Reed and others, all of which are invaluable books of reference for the home taxidermist.

It is to be regretted that the once flourishing Society of American Taxidermists has not been perpetuated, numbering, as it did, among its membership the best artists in their line in this country.

There is no royal road to success in this, more than any other of the arts and sciences, though I believe the ambitious beginner will find the way smoother; better materials are to be had, more helpful publications to be consulted and the lessening supply of wild life tends to make a more appreciative public than ever before.

CHAPTER II.

OUTFIT—TOOLS AND MATERIALS.

The extent and variety of work undertaken will determine the necessary working space and the assortment of tools needful. Other things being equal, the most complete assortment of tools and supplies makes possible the production of the best work in the minimum time. The equipment of the beginner need be but small and inexpensive, however, increasing the same as he discovers what is most necessary and desirable, in an increasing field of work. Wonderful pieces of taxidermy have been done with a pocket knife, pliers, needle and thread, some wire, tow and arsenic.

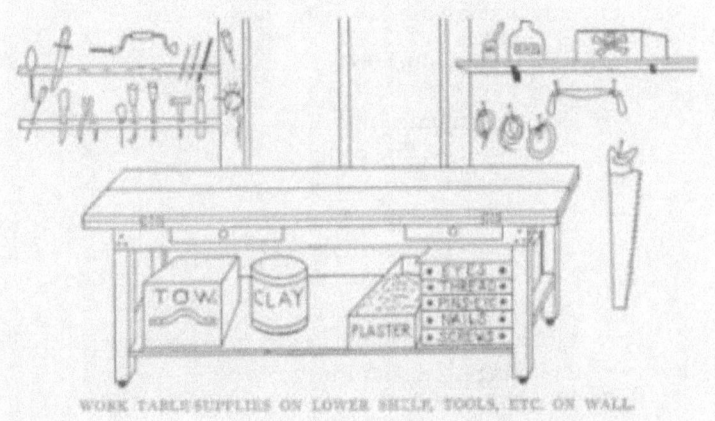

WORK TABLE SUPPLIES ON LOWER SHELF, TOOLS, ETC. ON WALL.

If no other room offers, much may be achieved (with the permission of the lady of the house) in a corner of a dining room or kitchen. A room or part of one well lighted, by north window or skylight preferably, makes the best location for the work table. This table may be of the common unpainted kitchen variety for all small work. It is well to make the top double by hinging on two leaves, which when extended will make it twice its usual width. When so extended and supported by swinging brackets it is specially adapted to sewing on rugs and robes. Such tables usually have one or two shallow drawers which are most useful to hold small tools. A shelf should be fitted between the legs, six or eight inches from the floor, forming a handy place for boxes of materials, books, etc.

If large work is in prospect a table should be built of the usual heighth, two or three feet wide, and six long. The legs of stout scantlings should be fitted with casters, making it easy to remove it to the center of the room where it can be approached on all sides, as will often be necessary. The double top, drawers, and shelf should be a part of the larger table also. Usually the table is kept in front of the window with tool racks and shelves for small articles each side of the same where they can easily be reached.

For preparing and mounting all small and medium size specimens I would advise the following list of tools. They will enable the worker to care for any of our native birds, quadrupeds up to the coyote, and any of our game heads, fur rugs, etc.

1 small skinning knife
1 medium skinning knife
1 larger skinning knife
1 pair scissors, fine points
1 pair shears, heavy, short
2 pairs flat nose pliers, large and small.
1 pair side or end cutters
1 pair fine forceps, 5 or 6 inch
2 flat files, large and small
1 adjustable tool handle, assorted tools, awls.
2 pinking irons, and 1 inch
Needles, assortment of cloth and glovers
Oilstone

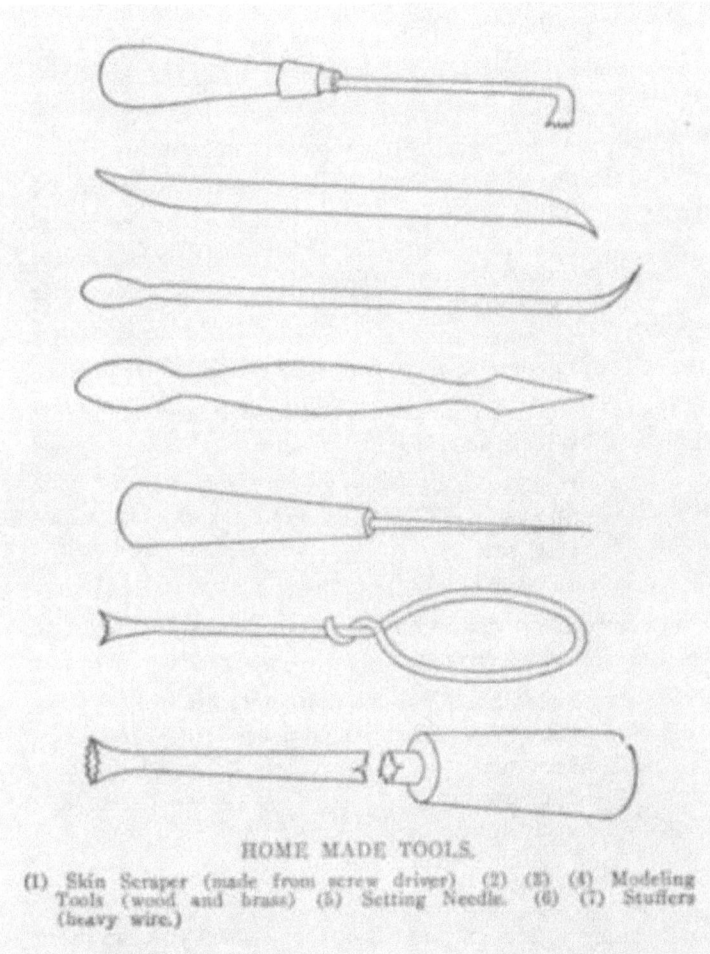

HOME MADE TOOLS.

(1) Skin Scraper (made from screw driver) (2) (3) (4) Modeling Tools (wood and brass) (5) Setting Needle. (6) (7) Stuffers (heavy wire.)

If it is capable of taking and holding a good edge the small blade of a pocket knife is equal to a surgeon's scalpel and a sharp shoe or paring knife, ground to the proper shape, is a nice medium size for skinning or trimming skins. A hunting or butcher knife is sufficient for the largest size. A few carpenter's tools are necessary and a complete set does not come amiss if much large work is attempted.

We must have:

Handsaw
Hatchet
Hammer
Bit brace
Assortment, drills and bits, in. and less.
Drawshave
Screwdriver
Small grindstone or corundum wheel
Chisels, two or three sizes
1 wood rasp
1 cabinet rasp
1 chopping block, made of a section of hardwood log

If large animals are to be mounted we will need in addition some iron working tools, such as

Set of taps and dies to 1 in.
Monkey wrench
Hack saw
Tanner's knives, 1 or more

A combination vise and anvil will be needed in any case as well as some miscellaneous tools:

Fur comb, coarse and fine combined
Paint, wax, and varnish brushes
Foot rule
Tape measure
Putty knife
Pointing trowel
Skin scraper

and some stuffing and modelling tools which you can make yourself. The list of materials seems like a long one, but many are inexpensive and others are used only in some small amounts, so the aggregate cost is small.

Excelsior
Fine tow
Cotton bat or wadding
Plaster paris
Corn meal

Gasoline
Potter's or modelling clay
Set tube oil colors
Glass eyes, assorted
Soft wire, assorted
Pins
Cord
Spool cotton, coarse and fine, black and white
Wax, varnish, glue, paste
Papier mache, or paper for same
 An assortment of nails, tacks, brads, screws, screw eyes and staples

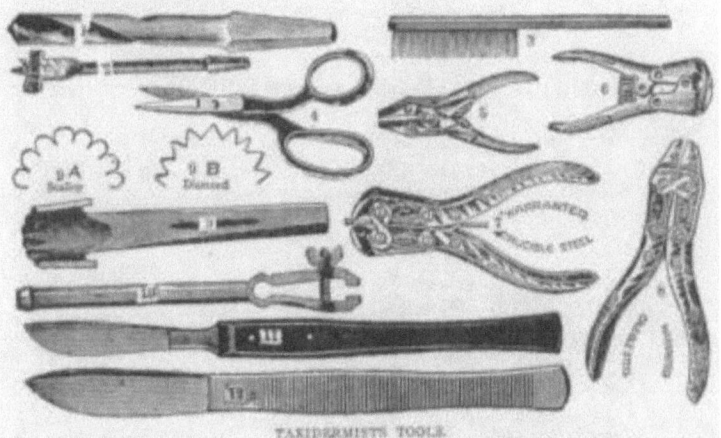

TAXIDERMISTS TOOLS.
(1) Chisel (2) Expansive bits (3) Brush (4) Bone cutter (5) (6) (7) (8) Pliers (9) Pinking irons (10) Hand vise (11) (12) Scalpels and knives.

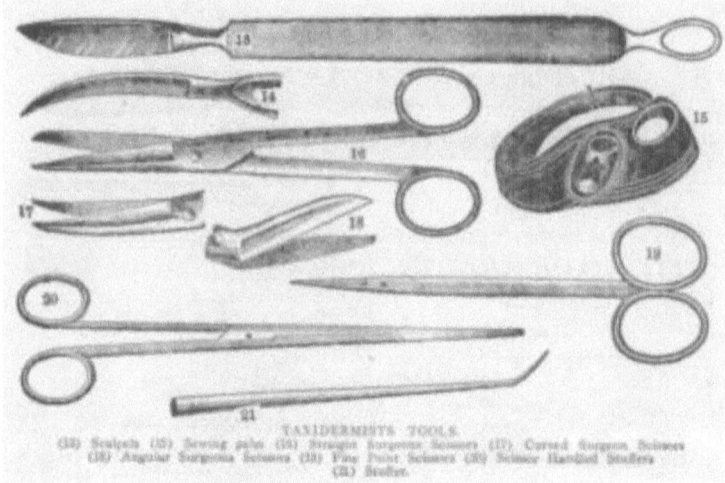

TAXIDERMISTS TOOLS.
(13) Scalpels (15) Sewing palm (16) Straight Surgeons Scissors (17) Curved Surgeons Scissors (18) Angular Surgeons Scissors (19) Fine Point Scissors (20) Scissor Handled Stuffers (21) Stuffer.

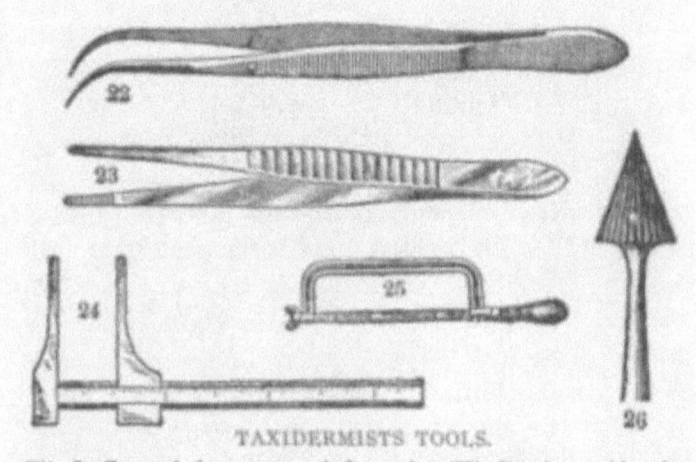

TAXIDERMISTS TOOLS.
(22) Stuffer and forcep, curved fine point (23) Regular taxidermists stuffer and forcep (24) Gauge (25) Hack Saw (26) Egg drill

A packing box or two will furnish some lumber for temporary stands and interior frame work. The permanent mounts are treated of elsewhere.

In ordering glass eyes it is often best to get them in the clear glass iris with black pupil so they may be given any color desired by painting the backs with tube colors, afterward protecting the paint with varnish. In this way a small stock will answer for many varieties. The plain black eye which is the least expensive can be used for many of the smaller birds and mammals, but should never be when the iris of the eye has any distinct tint. Do not make the mistake of ordering an assortment of "off" sizes and colors, that is those which are seldom called for. Aim to have those on hand for which you will have the most frequent use, the exceptions can be quickly had by parcel post. There is more demand for eyes of some shade of yellow or brown than any other colors, probably.

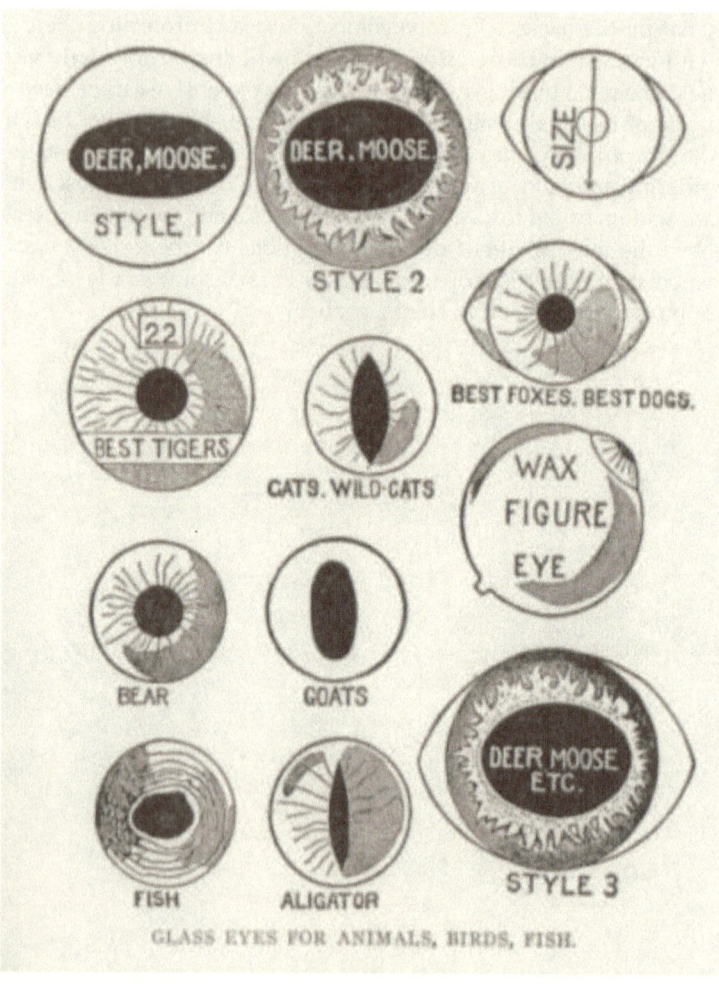

GLASS EYES FOR ANIMALS, BIRDS, FISH.

All birds have the round pupil. Elongated pupils are suitable for horned game and the cat tribe, irregular pupils fish, and the veined iris for dogs, wolves and foxes. Suitable sizes for some species of birds are as follows:

Number 3 to 5   Sparrows generally.
Number 5 to 7   Robin, blue jay, flicker.
Number 7 to 9   The smaller hawks and herons, nearly all the ducks.

Number 10 to 12 The smaller owls, the wild goose.
Number 12 to 14 The larger hawks and herons.
Number 15   Screech owl, eagles.
Number 17   Barred owl.
Number 19 or 20 Snowy owl and great horned owl.

Size of eyes for quadrupeds:

Number 7 or 8   Mink, skunk, red squirrel.
Number 10 to 12 Gray and fox squirrel, wood chuck, raccoon and opossum.
Number 12 to 14 Rabbit, small dogs, house cat.
Number 15 to 17 Jack rabbit, fox, medium size dogs, wild cat, black bear.
Number 18   Large dog, wolf, lynx, and grizzly bear.
Number 20 to 22 Puma, jaguar, small deer.
Number 23 to 24 Large deer, tiger.
Number 25 to 27 Moose, elk, caribou, horse, cow, lion.

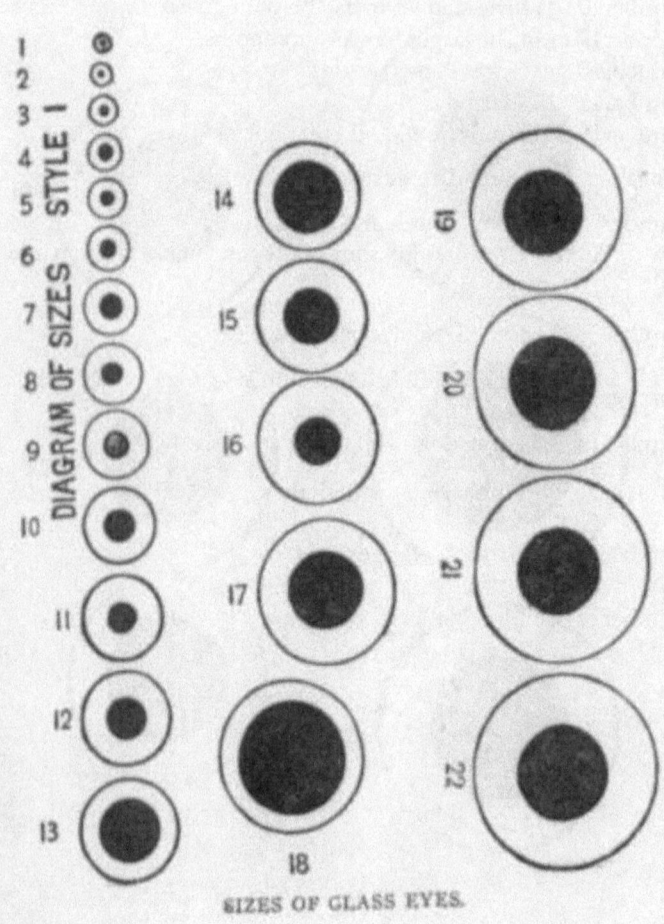

SIZES OF GLASS EYES.

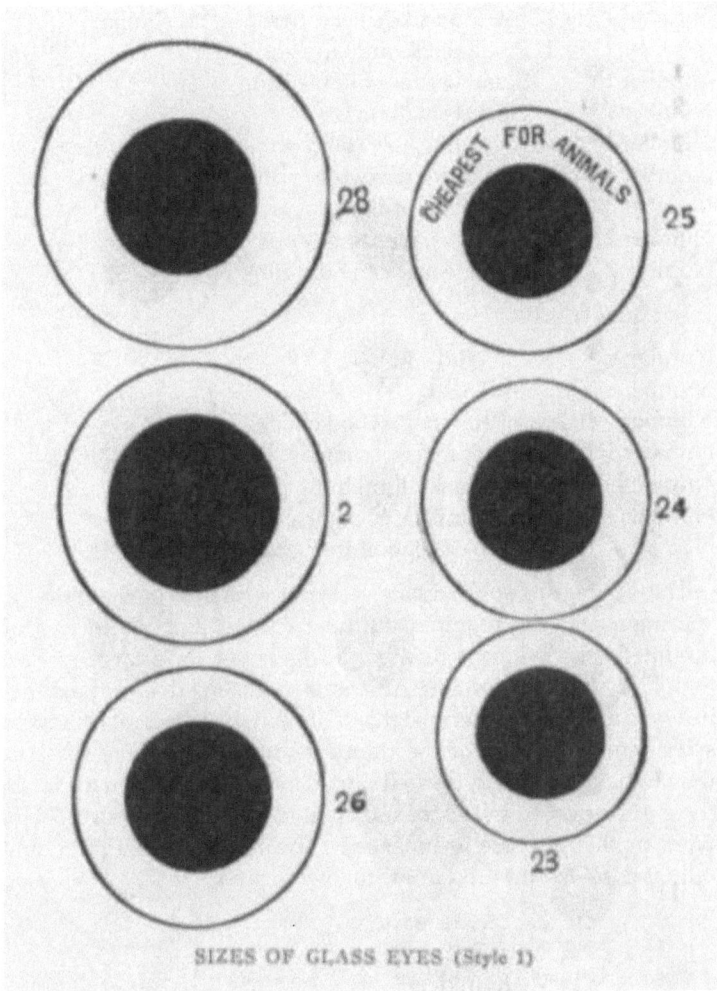

SIZES OF GLASS EYES (Style 1)

Of wire the following sizes are suitable for birds:

Number 6	Pelican.
Number 7	Wild turkey, swan.
Number 8	Eagles.
Number 9	Loon, goose, large herons.
Number 10	Seagull, large ducks, hawks, owls, and fish hawk.

Number 11 or 12 Medium size ducks, herons, and similar.
Number 13 or 14 Small ducks and grouse.
Number 15   Small herons, and medium owls.
Number 16   Doves, small owls.
Number 17 or 18 Bob white, jay, robin, snipe.
Number 19   Blackbird, waxwing, oriole.
Number 20   Bluebird, cardinal.
Number 21 to 24 Warblers, wrens, titmouse, finches.
Number 26   North American humming bird.

For quadrupeds:

Number	7	Wild cat.
Number	8	Foxes.
Number	10	Raccoon, wood chuck.
Number	11	Skunk, opossum.
Number	12 or 13	Muskrat, rabbit.
Number	14 or 15	Mink, large squirrels.
Number	17 or 18	Weasel, bull frog, and small squirrels.

These sizes are approximate, varying with size of the individual specimen and the required attitude wanted. For instance, a bird mounted with wings spread would be better for a large size wire than if in a resting position. An animal crouched does not require as heavy supports as one upright or in action. It is best to give the specimen the benefit of the doubt, as nothing is more disastrous than to have an otherwise well mounted subject sag down and spoil the entire effect from lack of sufficient mechanical support. The best wire for this purpose is annealed, galvanized iron. Larger animals require Norway iron rod in the following sizes:

in. Coyote, setter dog.
5/16 in. Wolf, puma.
3/8 in. Medium deer.
in. Caribou and large deer.
in. Moose, elk.

A large earthen jar or two will hold sufficient skin pickle for small animals. For large animals or great numbers of small ones a tank or barrel. Keep such jars or barrels covered to prevent evaporation. With dry arsenic and alum, arsenical solution, formaldehyde for an emergency and plenty of salt, even a beginner should be able to save almost anything that falls into his clutches.

There are numbers of reliable dealers in tools and supplies for the taxidermist and a perusal of their catalogs will be helpful, among the number being James P. Babbitt, 192 Hodges Ave., Taunton, Mass. Ready to use head forms, pinked rug and robe trimming, artificial tongues and ear forms, and even paper head and neck forms for the mounting of large game heads are some of the time and labor saving supplies they list. If you cannot attain to these, emergency supplies can be had of the dealer in hardware and dry goods, and one who cares for the art will rise superior to the material at his hand. What you "stuff 'em with" is of small consequence provided you use brains in the job. I have seen an elk head stuffed with old clothes with the bottoms of pop bottles for eyes, but would advise some other filling if possible.

CHAPTER III.

PRESERVATIVE PREPARATIONS, FORMULAS, ETC.

Probably arsenic in some form has been, and will continue to be the leading taxidermic preservative, or rather, insect deterrent. Many people are shy of handling this, but with reasonable care the use of arsenic is perfectly safe. Always keep poisons well labeled and out of the way of children. Nine children out of ten would never think of sampling them, but the tenth might prove the fatal exception.

There is far less danger to the operator in handling the needful amount of poisons than in endeavoring to save some rare but over-ripe subject. In many years' use of arsenic, dry, in wet solution, and in soap, I have received nothing more serious than an occasional sore finger.

The shape in which I have found it most satisfactory for poisoning hair and feathers of mounted specimens and the interior of furred skins I will give as

ARSENICAL SOLUTION.

Commercial arsenic	1 lb.
Bicarbonate of soda	lb.
Water	5 pts.

Boil until arsenic and soda have dissolved, stirring frequently. Use a vessel at least twice as large as necessary to contain the quantity used as it foams up while boiling. When cold put in a large bottle or jar marked *Poison*, of course. For poisoning finished specimens, mounted heads, etc., take one part of this solution to two parts water and spray the entire surface with this in an atomizer or larger sprayer. It should be tested before using by dipping a black feather in it and if a gray or white deposit is left on drying, it should be diluted still further until this is prevented.

To poison the inside of skins we make Arsenical Paste: Arsenical Solution (full strength), whiting sufficient to produce the consistency of cream. This should be mixed in a wide mouthed bottle or small pan and applied with a common paint brush. Do not apply to a perfectly dry skin, like tanned hide for a robe or rug, but dampen the inside first with clear water, then paint over with the paste and it will strike through to the fur side and be taken up around the fur roots by capillary action. This tends to put a damper on the activities of the moth, whose favorite grazing ground is at the hair roots just outside the skin.

The paste is equally good on skins of birds, except, perhaps the smaller ones, when freshly skinned, and some of the smaller mammals. The mixture of whiting makes it easy to see when the surface has already been treated, unless the skin is dressed white, in which case dry ochre may be used in place of whiting.

For poisoning the skins of the smaller furred animals and all but the larger birds:

DRYPRESERVATIVE.

Powdered white arsenic

Powdered alum

Mix equal parts by measure and apply to inside of the fresh skin with a soft brush or pad of cotton. If during the skinning and cleaning the skin has dried so the powder will not stick, moisten the inside of the skin with water before applying. Some taxidermists prefer to use in place of the paste some form of Arsenical Soap. This may be purchased from the supply dealer or made at home at quite a reduction. Personally I dislike the greasy, sticky feel of it; it is apt to cling around the finger nails and scratches, making them sore.

However, the following is the best formula for a time honored preservative:

HORNADAY'SARSENICALSOAP.

White bar soap	1 lb.
Powdered arsenic	1 lb.
Camphor gum	2 oz.
Subcarbonate of potash	3 oz.
Alcohol (wood or grain)	4 oz.

Slice the soap and melt slowly with a little water. When melted stir in the potash and arsenic. Boil to the consistency of molasses and pour into a jar to harden. Add the camphor already dissolved in the alcohol and stir occasionally while cooling. Mix with water and apply with a paint brush to flesh side of skins.

In case one is timid about using any of the arsenical preparations I would advise them to try

BROWNE'S NON-POISONOUS PRESERVATIVE SOAP.

Whiting	24 oz.
White soap	8 oz.
Chloride lime	oz.
Tincture camphor	1 oz.
Water	1 pt.

Shave the soap thin and boil it with the whiting and water till dissolved. Then remove from the fire and stir in the chloride, adding the tincture camphor later when cold, as much of the strength of the latter would be lost were the mixture hot.

Keep in closed jars or bottles, and if too thick, thin down with water or alcohol. Apply like arsenical soap or paste. This is highly recommended by English writers. For a non-poisonous powder I would advise equal parts of powdered alum and insect powder in place of the arsenic and alum.

SOLUTION FOR THE PRESERVATION OF SMALL SPECIMENS ENTIRE.

Formaldehyde (40% strength)	1	part
Water	9	parts

Mix thoroughly and put in glass or earthen jars or large mouthed bottles.

While most of the heads and skins of big game are preserved until they can reach the taxidermist, many of the smaller specimens become a total loss. Lack of time and knowledge are the chief causes of this loss of valuable souvenirs of days out of doors and interesting natural objects. Probably the easiest and least expensive method of temporarily preserving entire the smaller animals, birds, fish and reptiles is by immersion in the above preparation.

I would not advise its use on animals larger than a small fox or cat, and to insure an immediate penetration of the flesh the abdominal viscera should be removed from the larger specimens. The amount of solution used should be about ten times the volume of the subject, and it had best be replaced with fresh liquid after two or three days. I think this will work equally well on birds, reptiles and mammals.

On removal from the solution they may be skinned and mounted as fresh specimens. On removing from the solution, rinse in water containing a little ammonia to neutralize the irritating odor of the formaldehyde. Do not stand over the solution while mixing as the fumes of the formic acid affect the eyes. The condensed form in which this chemical can be carried and its cheapness (30c. per lb.), make it desirable as a temporary preservative. The saying, "It never rains but it pours," applies to the taxidermist and a sudden rush of subjects may often be saved by using the foregoing preparation. Other work may be under way, or for other reasons it may be desirable to keep a specimen in the flesh a short time before mounting.

ALCOHOLICSOLUTION.

Alcohol (94% strength)　　　　　　　　Equal parts
Water

If alcohol is less than 94% use less water. Use same as formaldehyde solution. This is said to be superior to the formaldehyde solution, though more expensive and harder to carry about on account of its greater bulk before mixing.

Specimens kept long in any liquid are apt to lose their colors. This fading will be reduced to a minimum if kept in the dark.

In order to do any satisfactory work on quadrupeds the taxidermist makes use of a bath or pickle of some sort for keeping skins in a wet state. This pickle sets the hair and in a measure tans the skin, reducing its liability to shrinkage and rendering it less desirable pasturage for insects.

All furred skins of any size should be immersed in this for a time before mounting, and may be kept in it for months or years without injury. If you have time to skin an animal properly the skin may be dropped in the pickle jar and in a day, week, or month be better fit for final mounting than at first.

For the first few days it is necessary to move it about every day so all parts may be exposed to the action of the pickle. The form of pickle which I have found most helpful is:

TANNINGLIQUOR.

Water	1	gal.
Salt	1	qt.
Sulphuric acid (measure)	1	oz.

Bring to boiling point to dissolve the salt. Allow to cool before stirring in the acid. When cold is ready for use.

When keeping skins in a wet state a long time I would prefer to use:

SALTANDALUMPICKLE.

Water	1	gal.
Salt	1	qt.
Alum	1	pt.

Boil to dissolve salt and alum; use like preceding.

If skins are to be kept some time they should, after a preliminary pickling, be put in new, fresh pickle and it should be occasionally tested with a salinometer and kept up to the original strength. Dirty and greasy pickle should be thrown away, but if clean and of low strength it can be brought up by adding new pickle of extra strength. It will do no hurt if more salt, even a saturated solution, is made of either of the foregoing.

While the salt and alum or acid pickle will keep our animal skins safely and in a relaxed condition ready for further preparation at any time, it will not answer for bird skins. For this we have a solution for keeping bird skins soft:

| Glycerine | 2 | parts |
| Carbolic acid | 1 | part |

After skinning the bird and applying some arsenical solution to the inside, brush this solution liberally over the entire inside of the skin. Pay special attention to the bones, wing and leg, skull and root of the tail. If necessary the skin may be packed flat for shipment. One treatment will keep all but the larger skins soft for several months. The feet, of course, will become hard and dry and must be relaxed as usual before mounting.

Every taxidermist needs to be more or less of a modeler, and one of the most useful materials is:

PAPIERMACHE.

Wet wood pulp	10 oz.
Glue (hot) or LePage's (measure)	3 oz.
Plaster paris, dry	20 oz.

This formula may be varied at the convenience of the operator. Work the glue into the pulp and knead the plaster into the mass. The more glue the slower it sets and a few drops of glycerine will keep it soft several days. Made with little or no glue it hardens quickly.

If the paper pulp is not at hand it may be made by tearing old newspapers or sheathing paper small and boiling and pounding till a pulp results. This composition is much in use in Europe in the making of many familiar toys and similar objects.

For modeling open mouths, finishing mounted specimens, making artificial rocks, stumps and boughs, it is very desirable.

WAXFOROPENMOUTHS.

Beeswax	1	part
Paraffin wax	1	part

Melt and color with tube oil colors. To color dip up a spoonful of melted wax, squeeze some tube color in and stir until stiff. Place the spoon in the hot wax and stir till evenly mixed. Do not try to put the color directly in the hot wax as it will not mix evenly so.

Wax should be melted in a water bath, like a glue pot, as excessive heat will darken it. Cakes of wax of suitable colors may be had of the supply dealers and are most economical when no great amount of work is done. The same parties supply the paper pulp previously mentioned.

CHAPTER IV.

PANELS, SHIELDS AND NATURAL AND ARTIFICIAL MOUNTS.

The preparation of a suitable setting for almost any mounted specimen will add greatly to its attractiveness. If you know where it is to be placed it is not difficult to make it suit its surrounding. For instance, a head of big game for hanging in a dining or ball room is suitably mounted on a polished and carved hardwood shield. While this would hardly match its surroundings on the wall of a log camp, a rustic panel of natural wood with the bark on would perfectly suit the latter place.

Heads, horns, and antlers are usually mounted on what are called shields. Fish and trophies of dead game birds and small game on panels. Single specimens are placed on severely plain wooden bases (museum style) or on those simulating branches, rocks, stumps or earth. These are artificial, but those built up in part at least with natural objects are most pleasing.

As we can not produce the best patterns of shields without special machines we must patronize either the supply dealers or the wood working mill for such. If convenient to a mill equipped with jigsaw and moulder they can be made up after your own patterns.

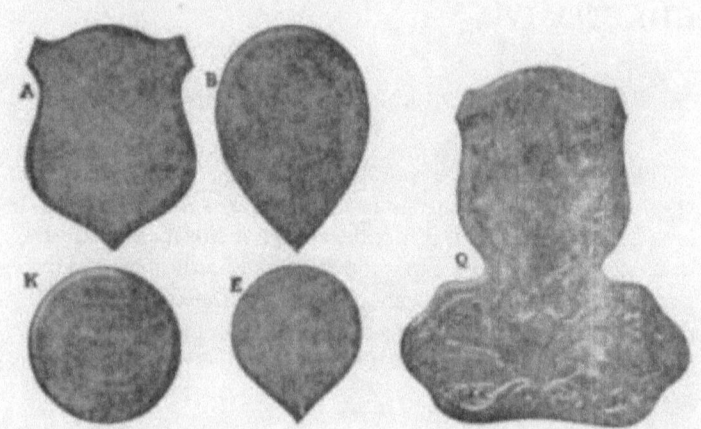

SHIELDS—VARIOUS KINDS AND SIZES.
(A) Suitable for moose, caribou, deer, fox by making or ordering according to size wanted; (B) Moose, caribou, deer; (K) Round shield; (E) Bear shield; (Q) Combination—head and feet.

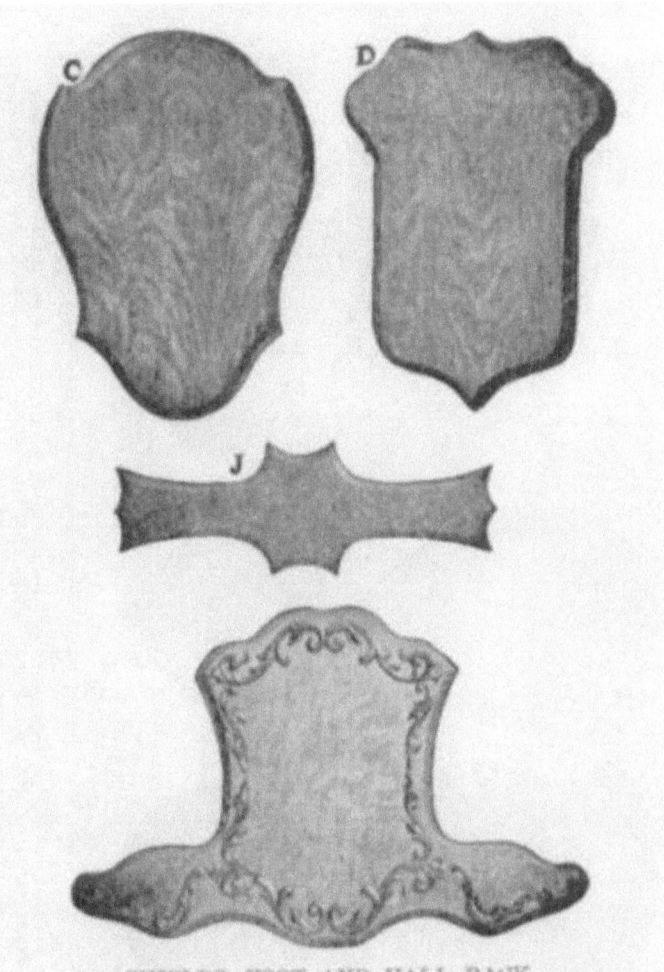

SHIELDS, FOOT AND HALL RACK
(C) and (D) Deer Shields (J) Four Feet Rack (V) Hall Rack

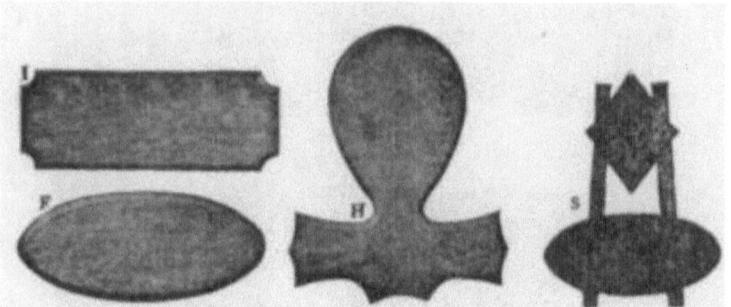

GUN RACK, FISH AND GAME PANELS, HALL RACK.
(I) Gun rack (F) Fish panel (H) Shield, combination head and feet (S) Hall Rack, small.

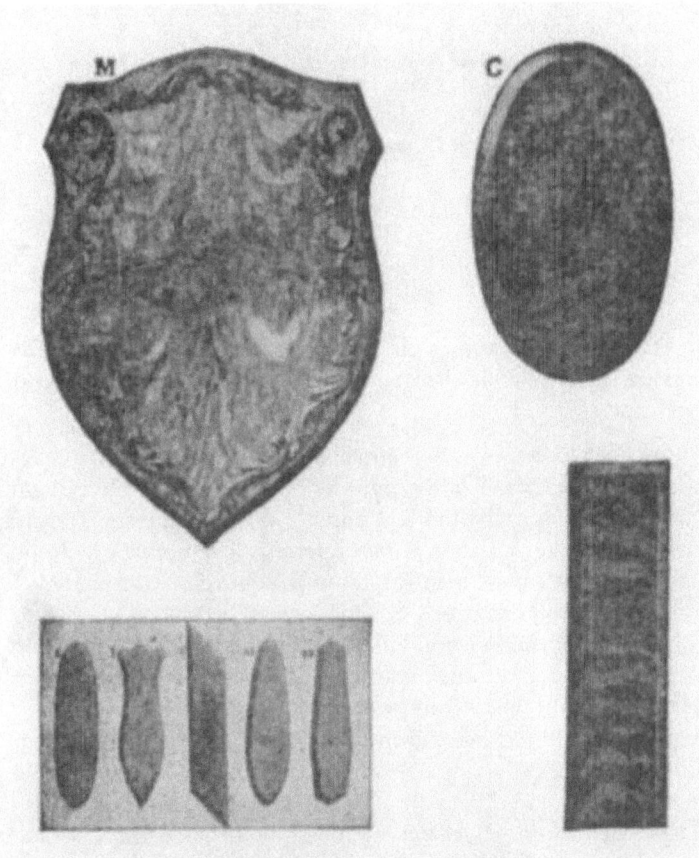

SOME SHIELDS AND PANELS.

(M) Shield with carving (G) Panel for game, x, y, z, aa, bb Deer foot thermometers (L) Deer foot thermometer and small animal panel.

Some of the sizes most used are approximately as follows for mounted heads:

For moose, elk, caribou.
20'0 inches.
For deer, goat or sheep.
12"8 or 16"1.

For fox or lynx.
8"0 inches.
For bear or wolf.
12"5
For birds, small furbearers and fish.
6☐ in.
Oval panels for mounting fish.
9"215"0
For dead game.
10"1514"417"5

For mounting horns of elk and moose the size for deer heads will answer nicely, while deer antlers are suitable with a shield of the fox head size.

In order to draw a symmetrical pattern for the woodworker, take a piece of stiff paper of the right length and width, fold it down the middle, draw one half to suit and cut out with shears. The style of moulding called Ogee is to be preferred. A simple diamond, heart, or oval shape can be made at home with beveled or rounded edges, or if your tools include a turning saw (which is most useful for a variety of purposes) you may try a more pretentious shield. To achieve this, make your pattern as just described and after marking it on a piece of wood from ⅜ to ⅞ inch thick, cut out with the turning saw. It should be held in the vise for this operation. Place this cut out shield (1) on a piece of board of similar thickness but somewhat larger and with a pair of compasses mark out another in. or so larger all around. (2) Also mark the same distance inside the edge.

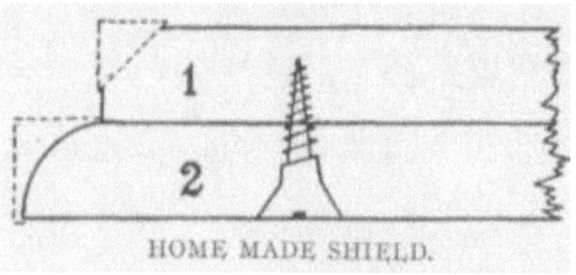
HOME MADE SHIELD.

With a wood worker's gauge or something similar make a mark around them both near the lower sides. Now with draw shave and rasp work the edges off both Nos. 1 and 2. No. 1 on a bevel, No. 2 rounded. There should be a number of holes drilled and countersunk in No. 2, from the back, and when the two parts of the shield are properly adjusted they should be drawn closely together with screws too short to penetrate the face of the shield.

If the adjustment is perfect the screws are to be drawn and the surfaces which come in contact coated well with glue, then drawn closely together and laid aside until thoroughly dry, when it should be well sandpapered before varnishing.

All shields and panels should be carefully sandpapered, filled and varnished, and polished if you wish. Don't make the shield or panel so ornate that the specimen will seem but an incidental, thrown in for good measure, so to speak.

Rustic panels can be made by sawing the end from a log on a slant, and planing smooth the oval. If this is heavily varnished on the front and back and the bark left on it is a very suitable mount for small heads, fish and birds. Artificial branches and trees for mounting birds should be avoided if possible; they are made by wrapping tow around wires, coating with glue and covering with moss or papier mache and painting. The result I consider unnatural and inartistic.

I would advise to use natural branches as far as possible; sufficient labor will be required to make necessary joining and changes look as near like nature as you can. Rock work is usually made of a wooden framework covered with cloth, wire or

SMALL BIRD.

paper and finished with a coating of glue and crushed stone or sand. One of the most useful materials in this work is the rough cork bark so much used by florists. It is light, comes in desirable shapes, can be nailed, sawed or coated with glue or paint. For constructing stumps for mounted birds of prey and rustic stands for small and medium fur animals it has no equal. Some taxidermists produce rock work of an obscure geological period by covering screen wire forms with a mixture of flour, baking powder and plaster of paris and water. This is put in an oven and baked hard, the weird result being painted to the artist's taste.

Water worn roots such as are found along the shore, twisted laurel branches, limbs of gum, oak and sassafras, all work up well in this and should be stored up to dry against a day of need. Out door people have a good eye for such things, but they are hard to find when you look for them, so gather them on your rambles. Papier mache is also a good modeling material for stumps, limbs and rock, being light, and readily taking coats of glue or paint. The expert can copy nature closely with it.

Some leaves and grasses can be pressed, dried and colored their natural shades with oil paints. The dealers supply a great variety of artificial foliage, some of which may be used to advantage, in case work especially. Dried mosses and lichens of various sorts may be used in this. Some of these powdered and glued on papier mache or cork bark stumps and limbs produce a very pleasing effect.

Snow scenes are frequently attempted but are not always a success. The peculiar fluffy and glittering appearance is rather difficult to reproduce. Torn or ground up white blotting paper mixed with a little ground mica has been used for this purpose. Glass icicles are listed by dealers and are quite natural in appearance, but the simulation of water is difficult and often disappointing.

It is often desirable to mount small specimens, of birds especially, in cases of some kind which will protect them from dust, dirt and rough handling and at the same time display them to advantage. The oldest and at the same time the least suitable contrivance for this is the well known bell glass or globe. It is difficult to find a safe place for this in the average house and it is not at all adapted to many specimens.

A plain wall case with glass front and a painted or decorated background will give the necessary protection with the least expense. For small bird groups, and singles and pairs of game birds, the oval convex glasses probably present the finest appearance. The backgrounds for these may be either plush or wood panels or hand painted, and any style of picture framing may be used. These are made in several sizes, listing at $2.00 to $8.00 each without backgrounds or frames. This cost has probably prevented their more common use.

There is on the market a papier mache background also adapted to any picture frame, called the "concave dust proof case." This has the flat face glass of the old style wall case, but with the square corners and much of the weight eliminated. Any of these styles of wall cases may be placed on shelves as well as hung on the wall like pictures, at once preventing breakage and becoming valuable decorations.

Special cases are often built (as in museums) for large and valuable mounted specimens. Of these the top and at least three sides should be of glass. The preparing and placing of the accessories in some large museum cases have required an unbelievable amount of time and expense to attain the desired natural appearance of the mounting.

CHAPTER V.

FIELD WORK, COLLECTING.

While it is unlikely that many readers of this book will undertake the collection of natural history specimens in any great numbers or as a special business, a few words on the subject may not be amiss.

It is well to bear in mind that the better the condition of the specimen when it first comes to hand, the greater will be our chances of success in properly preserving it. A small bird shot with a rifle is not worth bothering with unless excessively rare, and a fur bearer which the dogs have been allowed to maul and chew is very difficult to put in satisfactory condition.

One rule of the collector in the field is to shoot each specimen with the smallest possible charge of shot and powder which will kill it. I speak of shooting, as probably three-fourths of the objects mounted by the average taxidermist have been killed with fire arms.

Of late years a number of collector's guns have been put out by the arms makers, though any good small bore shotgun will answer for collecting all of our small and medium sized American birds and mammals. Some of these guns of about .44 cal. are exceedingly accurate and reliable performers.

In one case this small bore shotgun has been combined with a rifle, and the light weight and portability of this little arm makes it about the last word in guns for collecting all small specimens.

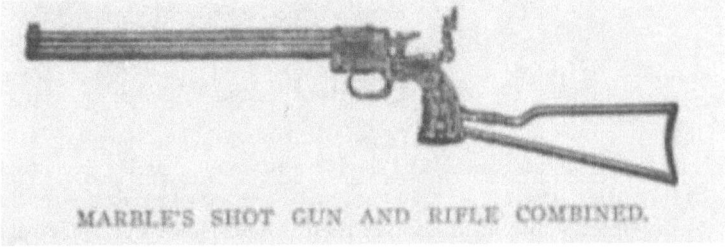

MARBLE'S SHOT GUN AND RIFLE COMBINED.

It as well as other single guns of the same bore, is built to use a round ball in the shot barrel, making them capable of stopping deer or bear at short ranges. However, choosing a gun is like choosing a wife, every one has their own tastes.

I would advise the would-be collector to load his own shotgun shells, at least those for small birds and animals, as it is almost impossible to get factory loaded shells but what are charged too heavily.

For the collection of animals for taxidermic purposes the use of traps will probably yield some of the best as well as the more rare and unusual varieties. Such styles of traps as least injure the appearance of the finished specimen are preferred.

The old-fashioned snare, dead-fall and box trap are as good as any in this respect. The wire spring or choke traps of Stop Thief style are ahead of the common steel trap in this respect, but like the homemade traps cannot be used in so many various situations.

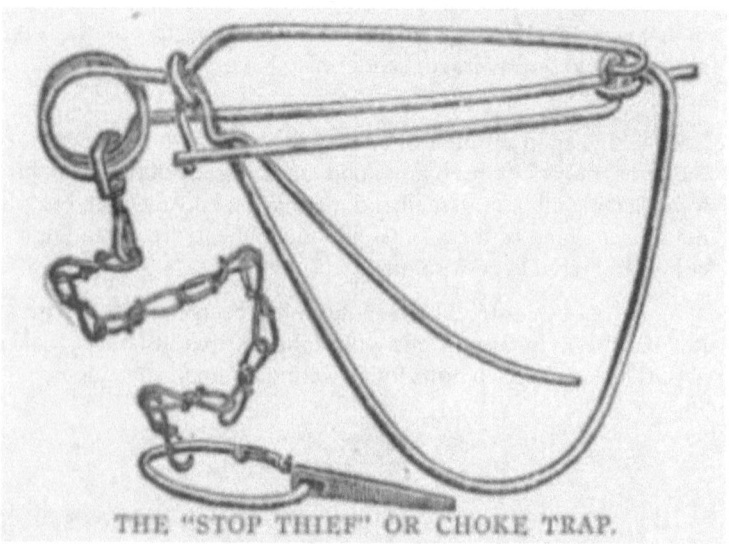

THE "STOP THIEF" OR CHOKE TRAP.

Water animals taken in steel traps may usually be quickly drowned. If set on land they should be tended often to prevent suffering and usually mutilation of the trapped game. Full information on this subject will be found in Science of Trapping and other books on special animals by same publisher.

The naturalist uses more small traps than large ones in most cases; many, many more specimens being taken in No. 1's than in bear traps. Several styles of mouse and rat traps are useful in collecting the smaller mammals, such as mice, rats and ground squirrels of various species.

Fish are to be collected any way you can get 'em, (legally at least). Many varieties of these are to be bought at the large markets and many rare and hideous specimens are discarded by market fishermen when culling their catches. A few years ago before much restriction was imposed on the sale of game it was possible to purchase many desirable things at the markets of Washington, D. C. Not only bear and deer, but elk, ptarmigan, arctic hares, sage and prairie grouse, fox squirrels, pileated woodpeckers and many other odds and ends were offered for sale as well as all the usual land and water game.

However you take your specimens or how badly damaged they may be when they reach your hands it behooves you to see that no further damage befalls them.

Specimens when shot should have all possible blood and dirt brushed or washed from feathers or fur and all shot holes, as well as mouth and nostrils plugged with a wisp of cotton to prevent further soiling. An awl, or piece of wire will be useful for this. Blood should be removed from white fur or feathers as soon as possible or it will be stained more or less. Small birds should be dropped head first into a paper cone, and laid in a basket or box if possible, the common hunting coat pocket is apt to break delicate feathers, though if the bird is well wrapped it may do.

Fur bearers will stand more rough usage and may be tied together by heads or feet or packed in game bag or pocket. Fish should be wrapped in paper to protect the scales.

It goes without saying that specimens which it is planned to preserve should be kept cool if possible until work can be started on them. Some varieties spoil more quickly than others; fish eating birds need quick attention; most birds of the hawk and owl family keep well, as do the pheasants, grouse, etc. Frozen animals keep perfectly in that state but spoil quickly after thawing.

Keep away from blow flies. Specimens are often sent to the taxidermists in apparent good order and when received are entirely ruined by fly maggots; the eggs being deposited before packing and shipping.

CHAPTER VI.

SKINNING AND PRESERVING SKINS.

With a suitable specimen at hand it is for us to decide if we shall mount it or preserve it as a skin temporarily or indefinitely. To illustrate we will presume that we have a muskrat just from the trap which is to be mounted at once.

Before skinning it is best to get some measurements to guide us in the later work. In this case where the skin is to be mounted immediately a simple outline is sufficient, as we will have the body in the flesh and all the leg bones, etc., to guide us in rebuilding the creature.

To get such outline, lay the animal on its side on a piece of blank paper, put the feet and legs in some natural position, fasten them in place with a few pins and mark around the entire animal with a pencil. The eye, hip and shoulder joints, and base of skull may be indicated on this outline sheet. Our muskrat is a trapped and drowned one so we will not have to replace the shot hole plugs with fresh ones, as would be best if it had been killed with the gun. Also it has been dead long enough for the rigor mortis to prevent the free flow of blood and body juices which bother the operator if it has been killed but an hour or less.

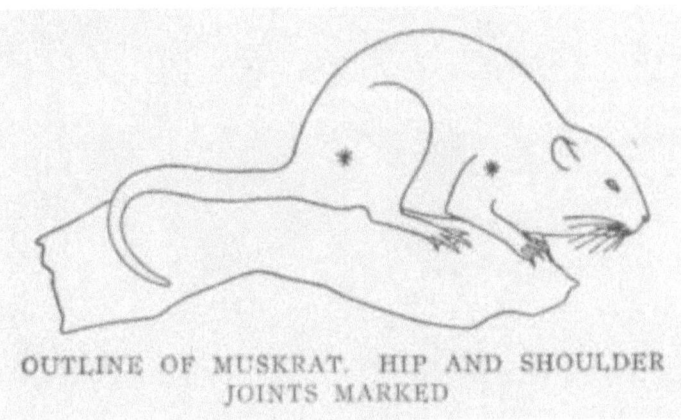

OUTLINE OF MUSKRAT. HIP AND SHOULDER JOINTS MARKED

Laying the animal on its back, make the opening cut by pushing the knife point through the skin at the juncture of neck and chest. Run the blade down between skin and flesh, separating the skin in a long clean cut to the root of the tail. Open the tail also along the under side from the tip to within an inch or so of its base. Slit open the sole of each foot from the middle toe to the heel and further if necessary so the leg skin may be turned down over the foot.

Beginning at this central cut, skin back each side until the shoulder and hip joints are encountered. Bending the limb will show the exact joint where the muscles are cut apart and the legs severed from the body. Cut off the tail near its base, leaving it in the skin for the present. Loosen the skin from back and shoulders and turn it wrong side out over the head. Skin down until the ears are reached, cutting them off close down to the head and continue on to the eyes. Work carefully around these and cut close to the skull to avoid hacking the eyelids. Cut through the nose cartilage, and when the lips are reached cut them away close to the gums, leaving both their inner and outer skin on the pelt. Cutting them off at the edge of the hair is a frequent cause of trouble as they are full and fleshy and should be split, pared down on the inside and when mounted, filled out to their natural shape to perfect the anatomy.

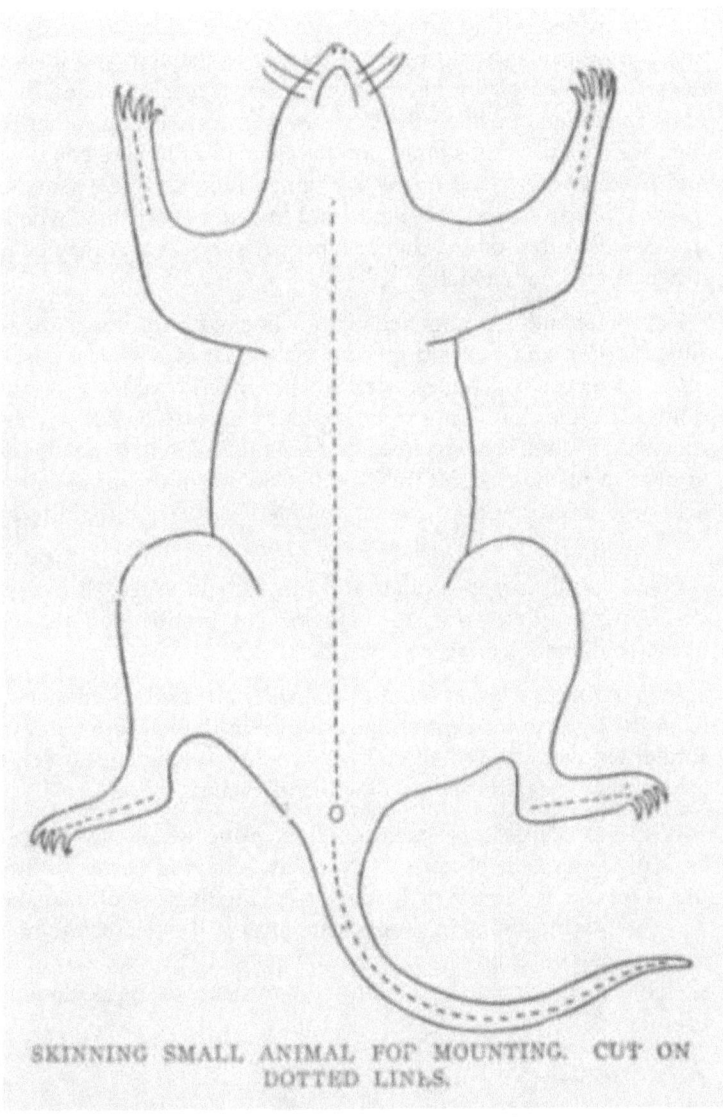

SKINNING SMALL ANIMAL FOR MOUNTING. CUT ON DOTTED LINES.

Now the skin of each leg may be turned wrong side out down to the toes and all flesh and muscle cut away from the leg bones with knife and scissors. The tail also is to be removed from its skin and the ears turned wrong side out to their tips. In skinning no flesh of any consequence should be left adhering to the skin and it should never be pulled off by main strength, but rather separated by the knife or fingers, pushing the flesh away from the skin rather than the reverse. The skull should now be cleaned as the leg bones were and if a number of specimens are being prepared at the same time the heads may be put in a kettle and boiled a short time, when on trying with a fork or awl the flesh becomes tender and may be rapidly removed with a knife.

The brain may be removed with a hooked wire, the skull well rinsed inside and out and given a good coat of arsenical paste or other preservative. The next step is determined by what you intend doing with the skin. If it is to be kept for purposes of study without mounting it should be made up as a scientific skin. If to be mounted at once or in the near future it should be put in the jar of salt and acid or alum. It can of course be mounted at once without this bath but I believe it is well worth any extra work it entails.

Some young furred animals and others with very delicate skins do not work up well from the bath and had best be mounted without it, being handled more as bird skins are.

In skinning the larger quadrupeds we make two additional cuts, from the right to the central line and out to the left fore foot and a similar cut connecting the hind feet. These opening cuts are on the back and inside of the legs, you will understand.

With most large subjects another cut from the shoulders up the back of the neck is necessary. On animals having horns or antlers this is terminated in a Y or T shape, reaching the base of the antlers. After loosening the skin around the antlers the head is removed through this incision. As it is hardly possible to make an outline sketch of a large animal, as full a set of measurements as possible is useful in all cases.

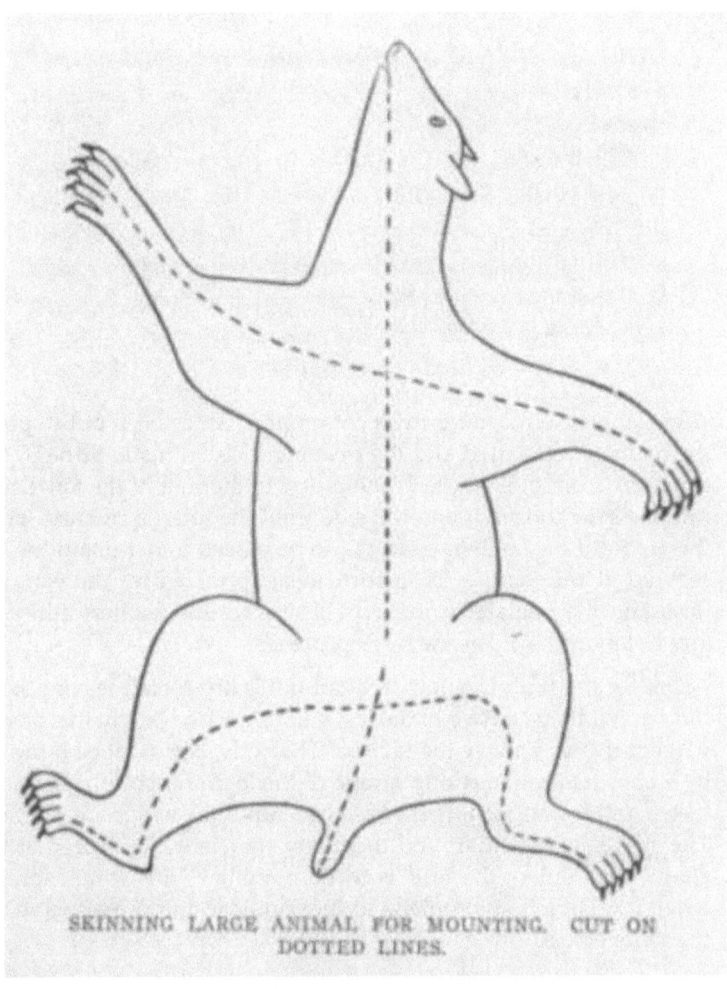

SKINNING LARGE ANIMAL FOR MOUNTING. CUT ON DOTTED LINES.

The bones of the legs will be in the way attached to the skin and the two upper bones of each leg may be detached and if lack of space or transportation make it necessary, thrown away, though if the bones of one hind leg and one front leg are preserved artificial duplicates may be carved.

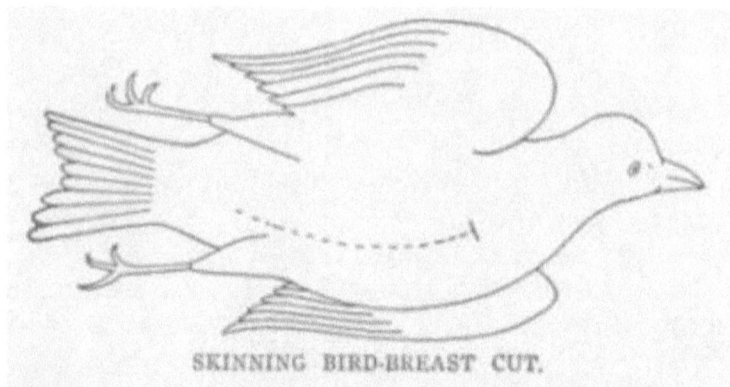

SKINNING BIRD-BREAST CUT.

In skinning birds, after fresh cotton plugs have been put in place the feathers are parted and the opening incision made through the skin only from the middle of the breast to the root of the tail. Separate the skin and flesh on each side until the knee is reached, push this up until the knife or scissors can be passed under it and the leg severed at the joint. A little corn meal sprinkled on the exposed flesh and the operator's fingers will prevent the feathers adhering and becoming soiled as the work proceeds.

Cut off the flesh in which the tail quills are rooted leaving it on the skin with one or two of the last vertebrae. Use care in this or you will cut the skin above the tail too. The body may now be hung up by a cord tied to the stump of one of the legs and both hands used in separating and turning the skin back until the wings are reached. The skin is loosened around these and they may be severed at the elbow joint unless the bird is to be mounted with wings spread, when it will be best to unjoint at the shoulder and preserve the entire wing bones.

With the wings detached we skin on to the base of the skull. In some of the ducks, and other water birds, woodpeckers and owls the neck is so slen-

HOODED MERGANSER.

der and the skull so large that it is necessary to cut the neck off here and making a cut down the back of the head and neck continue the skinning of the head through it. Do not cut or tear the membrane of the ears but pull it out with the forceps and work down over the eyes, cutting the membrane connecting the skin but not the lids or eyeball itself. The liquid contents of the eye are particularly sticky and difficult to remove from feathers.

Continue skinning to the base of the bill, scoop the eyes from their sockets and cut loose the forward part of the skull from the neck. This is usually accomplished with four snips of the scissors much easier to practice than to describe.

Make one cut on each side of the head, through the small bone connecting the base of the lower jaw with the skull, another through the roof of the mouth at the base of the upper mandible and between the jaws of the lower, and the last through the skull behind the eyes and parallel with the roof of the mouth. This will free the skull of the neck and most of its flesh and muscle.

In most cases the head should be returned to the skin as soon as possible to avoid its drying out of shape and giving the feathers a wrong set. After cleaning and poisoning the skull and filling the eye sockets with cotton this reversing is undertaken. If working on a small bird the learner is apt to come to grief here, as only by careful and patient work without the application of some force is the returning process accomplished successfully.

The wings and legs may now be skinned down to the first joint and all flesh and muscle removed from the bones. This is done expeditiously by snipping off the end of the leg bone and stripping it down with adhering flesh to the ankle joint where it (the flesh) is cut off.

The wings are skinned to the first joint, stripping the wing primary feathers from their fastening on the bone with the thumb nail, clipping off the large bone near the end and detaching the small bone with all flesh and muscle adhering. If this is clipped off at the wrist joint the entire wing is cleaned. This method applies to all small and medium birds not wanted with spread wings.

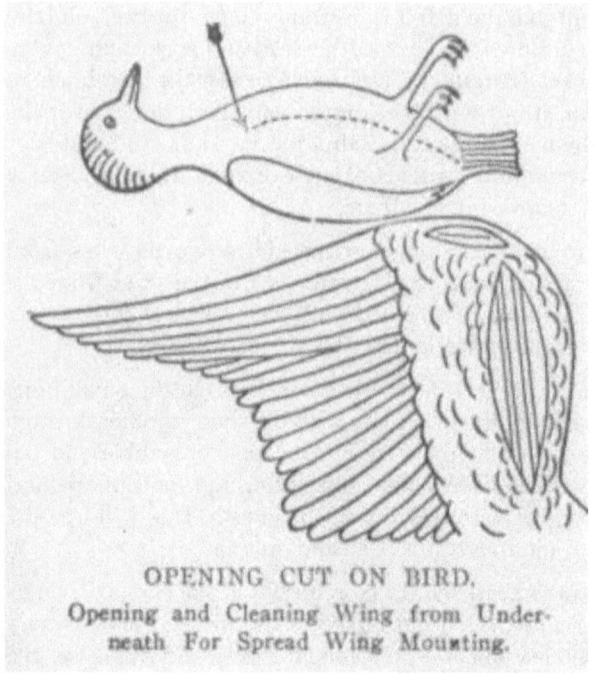

OPENING CUT ON BIRD.
Opening and Cleaning Wing from Underneath For Spread Wing Mounting.

In case wing spread is wanted the primary feathers are not disturbed but that part of the wing is cleaned from a slit in the under side of the wing. All but the smallest birds should have the tip joint of the wing slit open on the under side and some form of preservative worked in the opening. Specimens the size of a crow and larger should have a cut made in the bottom of the foot and the tendons of the lower leg drawn out with an awl, and in the case of very large birds it may be necessary to soak the unfeathered part of the legs and feet in a pan of strong pickle for 24 hours, to prevent decay and damage from insects. Our bird is now entirely ready for the application of such preservatives as we may be using.

The main principle in the preserving of skins may be stated thus: All skins must be removed and cleaned of flesh so the preservative may be applied to every part of the inner surface, where it will act directly on the roots of the hair or feathers.

The preservative applied, we must decide on the next step, whether it is to be mounted at once, in a short time, or laid away indefinitely as a scientific skin.

If we have to lay it away until tomorrow, put a little cotton inside to prevent the inner surface sticking together, wrap in a damp cloth and unless the weather is very hot it will be all right.

If very hot or it is necessary to lay it aside for some days, the inner side should be well coated with the carbolic and glycerine mixture. As a scientific skin it should be made up at once, tagged with a full set of measurements taken before skinning and laid aside to dry. These measurements are not needed if we mount it at once, as the skinned body is at hand for comparison, but the sex, date, locality and collector's name should be attached to the completed specimen.

Alligators and the lizard family are skinned like the other four-footed species, as are snakes and fish, with the exception they have no limb bones to be cleaned and preserved. Fish are better opened along one side than the central line of the body.

Reptile skins if not put in the pickle jar had better be packed in salt after poisoning as when entirely dry they are practically ruined. Skins of fish are best kept in either a saturated solution of salt (water with salt added until no more will dissolve), alcohol or formaldehyde solution. Whatever method is used the delicate colors will vanish and unless you can have a fresh specimen at hand when mounting it you should make the best color record you are able. This is true to some extent at least of all coverings of fur, feathers, or scales, and the stronger the light the more damage. I have seen a mounted mink placed in direct sunshine, bleached to a drab and the yellow feathers on a 'flicker faded almost white.

In order to preserve turtles, after killing with chloroform preferably, it is necessary in the case of the box or land turtles to cut a square opening in the under shell through which the body may be removed and the legs and neck skinned.

The water species can have the lower shell detached from the upper at the side and after cutting the skin around the rear two-thirds this shell is turned over to the front and the skinning and cleaning proceeded with.

If not mounted at once make into a dry skin after poisoning or small turtles may be put in alcohol.

CHAPTER VII.

MAKING SCIENTIFIC SKINS.

Probably most bird skins which are not mounted at once are kept in the form of "scientific skins." In other words they are skinned, poisoned and without wiring, given the shape of the dead bird. Their plumage, size, etc., may be examined, they are easily packed or shipped and, if properly made, may be mounted at any time but at the expenditure of considerably more work than a freshly taken skin requires.

The instructions on skinning leaves us with the skin wrong side out with the exception of possibly the head. The leg and wing bone, cleaned of flesh, should be well poisoned as well as the skin and after wrapping with pieces of cotton bat to their approximate size, returned to their places. It is well when doing this to under rather than over fill.

Connect the bones of the wings with a bit of thread or cord. After filling out around the eyes and upper throat, wind a small stick or piece of wire with cotton to a size a little smaller than the natural neck and push it into the opening at the back of skull.

The body can be filled out now with raw cotton, tow, or any similar substance not of animal origin. Fine excelsior is about right for large birds. The edges of the opening cut may be drawn together by a few coarse stitches. After the feet have been tied together it is time to adjust all the feathers become well rumpled in handling.

Our fingers, forceps and a setting needle made of a large needle or part of a hat pin in a wooden handle will accomplish this. Stained or dirty plumage should be cleaned before the skin is filled out, by first sponging with tepid water, then with gasoline or benzine and drying with plaster of paris or corn meal. Never apply this without the gasoline first or you will have trouble indeed.

Now the skin is ready for its label, which should supply the following information:

1. Length in inches from tip of tail to end of tail.

2. Distance between the tips of outstretched wings.
3. Length of wing from the first joint.
4. Color of eyes, feet, bill, etc.
5. Date, locality, collector.
6. The sex.

The first three items are often combined, in the case of a bluebird for instance, 7-12-4, the order being understood.

Unless the plumage plainly indicates the sex this should be assured by examination of the skinned body. By making an opening in the side of this near the back bone the inside surface of the small of the back is exposed. In the case of the male there will be visible two rounded bodies, varying in size with the season and species, and in the female a flattened mass of spheres.

After labeling and fastening the bill together with a pin or thread the skin should be slipped inside a paper tube to dry. Water birds with long slender necks should have the head bent around beside the body and the long legs of waders are bent at the ankle and left resting on the body; this to prevent breakage.

Duck, geese and any fat birds need the inside of the skins well scraped, sponged with gasoline, partly filled with plaster paris and left for several hours so all grease may be absorbed. This grease should be removed prior to applying the preservative as it will prevent any effectual penetration by the latter.

After cleaning either the inside or out of a skin with plaster it will be necessary to gently beat it with a whisk broom or something similar to dislodge the particles of plaster. A current of air (from a bicycle pump, for instance) will remove the dust from the feathers when dry.

SCIENTIFIC SKINS, SMALL ANIMALS AND BIRDS.

Fewest dry scientific skins are made up from the quadrupeds, but in case the matter of transportation prevents wet preservation or they are wanted dry the all around taxidermist must practice at making them up also. Like the bird skin they should be thoroughly rid of flesh and fat after skinning but do not require such finical handling. Rinsing in water with a little washing powder or soda added will remove blood stains and some grease but the benzine bath with the drying after, as recommended in the chapter on tanning, etc., will be needed in case of very fat specimens.

All small animals are made up about the same as birds, wrapping the leg bones in tow, oakum or cotton and filling out the body with the same material. The skull cleaned and poisoned had best be put in the centre of the body with the filling, when it can be found at any time by ripping a few of the stitches.

The skin of the head is filled out with the same material and the tail may either be bent up under the body or drawn together by a few stitches around a wrapped wire extending into the body half its length. Of course the operator will see that the entire inner surface of the skin is treated liberally with some preservative, arsenical paste preferably, before the filling process.

After stitching up the opening cut the skin is laid on a board, back up and the legs neatly disposed, the front feet beside the head and the hind ones drawn back beside the tail. The feet are fastened with a pin each and after smoothing down the fur with a small metal fur comb the skin is laid aside in an airy, shady place until fully dry.

With each scientific skin a record should be made of the following details:

1. Length, end of nose to root of tail.
2. Length of tail from root to end of bone.
3. Height at shoulders.
4. Color of eyes, lips, feet, etc.
5. Name of species, sex, locality, date, and collector's name.

These may be noted down on a corner of the outline sheet, which is numbered and filed away; the skin tagged with a duplicate number is put in the pickle jar or made up as a dried skin, whichever is desired, or the full information may be put on a tag attached to the skin. Many collectors simply number all specimens and preserve all information in their note books. The foregoing details are sufficient for animals less than bear and deer in size.

The larger animals should have as many as possible of the following additional measurements:

Distance hip joint to shoulder joint.

Circumference	of	forearm.
"	"	neck.
"	"	body.

Back of leg.

Weight if possible.

Skins of large animals, a bear for instance, may have a slight wrapping of tow or excelsior on the leg bones to prevent their coming in contact with the skin and the whole skin laid to dry on a scaffold of poles or something similar. When nearly dry fold up with the legs inside in a square shaped package. This can be tied up with heavy cord or even sewed up in burlap to prevent damaging the skin in transit. Fish and reptiles are not a success as dry skins.

CHAPTER VIII.

PREPARING DRY AND WET SKINS FOR MOUNTING.

Let us assume that we have a dry skin each of a small bird and a furred animal which has been properly made up sometime in the past and which it is necessary to mount. Taking the bird skin first, the usual way is to first wrap the unfeathered parts of legs in some strips of cotton cloth saturated with water containing a few drops of carbolic acid until they begin to relax or lose their stiffness somewhat.

Then the filling may be removed from the whole interior of the skin and be replaced with pieces of cotton, dampened as before, and the whole skin wrapped in a cloth or shut in a close box until with some scraping and manipulation it becomes as pliable as when first removed. Any little lumps of dried muscle should be broken up and the edges of the opening cut, scraped and stretched out as they are very apt to wrinkle and curl up, thus reducing the size of the skin considerably.

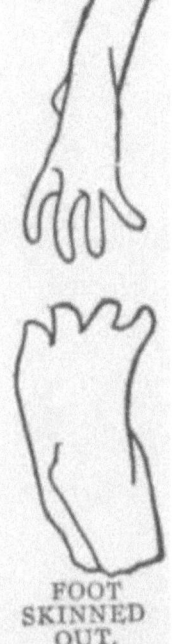

FOOT SKINNED OUT.

The eye sockets are to be filled with balls of wet cotton to render the lids and surrounding skin soft. The roots of quills and tufts of large feathers will need loosening as some flesh is necessarily left around them.

The small animal skin may be treated the same way but the most thorough and expeditious method of relaxing skins of both animals and birds (except the smallest of the latter) is to plunge them into water, clear in cool weather, slightly carbolized in warm, until they are pretty well relaxed. Then go after the inner side with scraper until any lumps of fat, muscle and the inner skin are well scratched up. Soak in benzine or gasoline and clean with hot meal, sand, sawdust or plaster as directed for tanning. Remember that bird skins must be handled carefully, so do not be too strenuous in beating and shaking them.

Of course if any skin has been laid away with quantities of fat adhering it will need very gingerly handling to save it, in fact unless *very* rare such skins are not worth trying to save as they have little durability however treated. The largest polar bear skin I ever saw was ruined by lying "in the grease" too long before dressing. Bird skins preserved with the glycerine carbolic preparation require relaxing the legs and a cleaning and dampening up of the inside of the skins.

Furred skins from the pickle need a good scraping on their inner surface, thorough rinsing in soda solution to neutralize the acid and remove all salt, then the benzine bath and cleaning. Don't forget to rinse salted or pickled skins else beads of moisture will form on the specimen in damp weather and crystals of salt in dry.

Occasionally an extra rare skin will drop to pieces through age or other infirmities when being prepared for mounting. The only hope for it then is to glue and pin it piecemeal on a manikin covered with some preparation which gives it a firm surface. While an expert will achieve fair results in such work the amateur could hardly expect success.

CHAPTER IX.

MOUNTING SMALL AND MEDIUM BIRDS.

A word of advice to the beginner as to the variety of specimen to use in first trials. Don't begin on too small a bird until somewhat adept; unpracticed fingers bungle sadly over tiny feathered bodies. A first subject should be at least as large as a bob white to give room to work, and of some variety in which the feathers are firmly embedded.

Snow birds, cardinals, and some others have very thin delicate skins, the pigeons shed their feathers on little or no provocation. Blackbirds and jays are very good to practice on but the very best would be a coot, sometimes called crow duck or mudhen. It is of fair size, closely covered with feathers which will fall in place readily after skinning and wiring even at the hands of a beginner.

Many, in fact most, birds have numerous bare patches which the adjacent feathered tracts cover perfectly while in the flesh, but which a too generous filling will exhibit in all their nakedness. I had not discovered this until some of my first attempts at mounting birds nonplussed me by showing numerous patches of bare skin in spite of the fact that but a few feathers had become loosened in the handling.

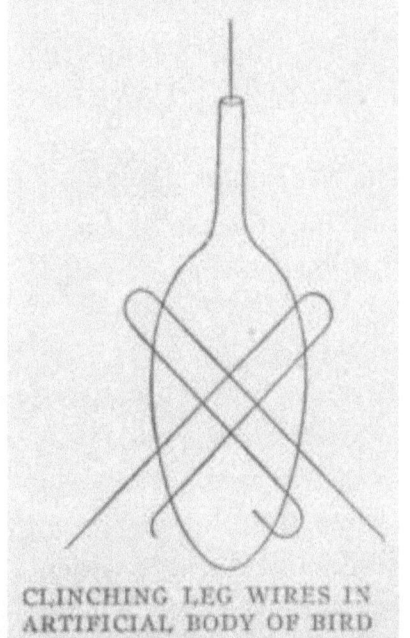

CLINCHING LEG WIRES IN ARTIFICIAL BODY OF BIRD

We will assume that a suitable specimen is at hand, freshly killed and properly skinned as per the directions already given. All bones remaining with the skin, lower leg, wing, skull, etc., have been stripped of flesh and any shreds remaining poisoned, as has the entire inner surface of skin. With the skinned body at hand cut three wires of suitable size, one a little more than twice the length of the body and neck, for the body wire, the other two about twice the length of the legs may be a size larger as it is important that the leg wires furnish adequate support.

Form the body wire into a loop which is the outline of the body laid on one side with the surplus end projecting along the line of the neck. This loop should not be quite as large as the body, however, to allow for a thin layer of filling material over it. Wad up a handful of coarse tow, push it inside the body loop and wind with coarse thread, drawing in by pressure and winding and building out with flakes of tow to a rough shape of the skinned body. The neck also is built up the same way, making it fully as thick as the original but no longer ever.

If the wire projects more than a couple of inches from this artificial neck, cut it off at that length and with a flat file or emery wheel give it a sharp triangular point. The leg wires, too, should be pointed similarly. All wire should be smooth, straight, and free from kinks to work well. Coming in coils it will require straightening, the larger sizes with mallet or hammer and No. 18 and smaller by fastening one end in the vise and giving the other a sharp tug with a pair of pliers. It will be felt to stretch slightly and become quite straight.

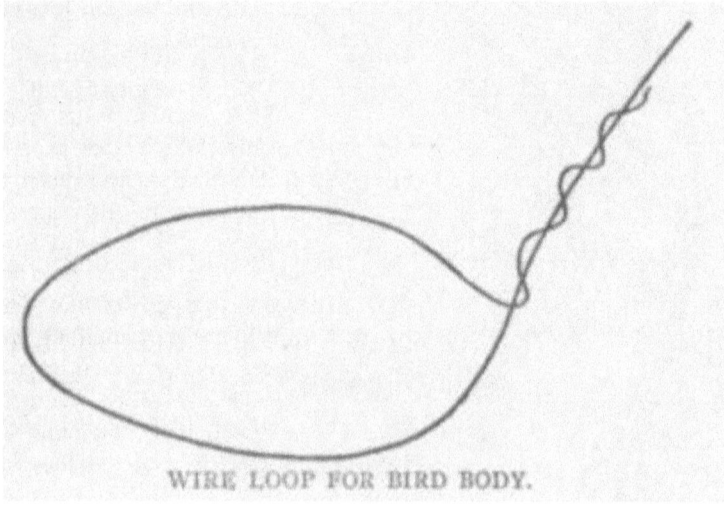

WIRE LOOP FOR BIRD BODY.

Next insert the pointed end of a leg wire in the bottom of the foot and pass it up along the back of the bone between it and the skin. A considerable knack is necessary to do this successfully and some force must be used. Passing the heel joint is difficult but having done this and emerged inside the skin continue to pass it until it is a little longer than the leg bone beside it.

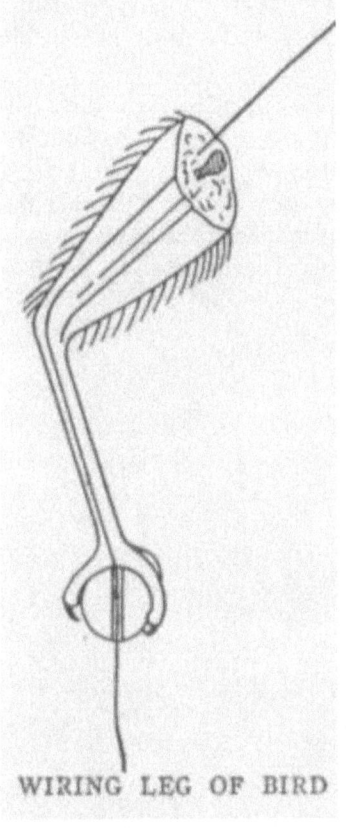

WIRING LEG OF BIRD

Turn the skin of leg inside out and wrapping tow around both bone and wire build up a duplicate of the leg from thigh to heel, wrapping snugly with thread. Treat the other leg the same.

With forceps fill the skull back of mouth with tow cut in short pieces. A quantity of this may be chopped on the block with the hatchet and kept on hand in a box. Never fill any part of a mounted bird with cotton unless it may be the sockets as it is impossible to force a sharpened wire or pin through it.

The parts of the wing bones remaining should be wrapped with tow as the legs are, only they hardly need any wiring inside unless the bird is to be with wings spread. Fasten the ends of the wing bones together by a stout cord or thread so they are separated the distance between the shoulders, measuring across the back of body. Now insert the neck wire in the back of skull forcing it out through the crown until the artificial neck is brought snugly against the opening at the base of the skull.

Bend the pointed end over to get it out of the way and adjust the skin of the neck. Draw the skin of breast over the body keeping the bird on its back. If the body has been properly made it will fill the skin rather loosely. If too large it can be removed and made smaller before proceeding.

The operator will note that in all small and medium birds the thigh and the upper wing, next the shoulder are not built up and wired with the rest of the limbs but are filled out later from inside the skin, as in all ordinary positions they show but little externally, the elbow and knee joints nestling close to the body among the feathers.

So when fastening the legs to the body let the wires enter where the knee would lie and push the wire through obliquely, upward and forward, pushing and drawing them through the artificial and natural leg until the lower ends approach the feet. Grasping the sharpened ends of the leg wires at the middle of the length projecting from the body, with round nose pliers bend them over in a hair pin shape.

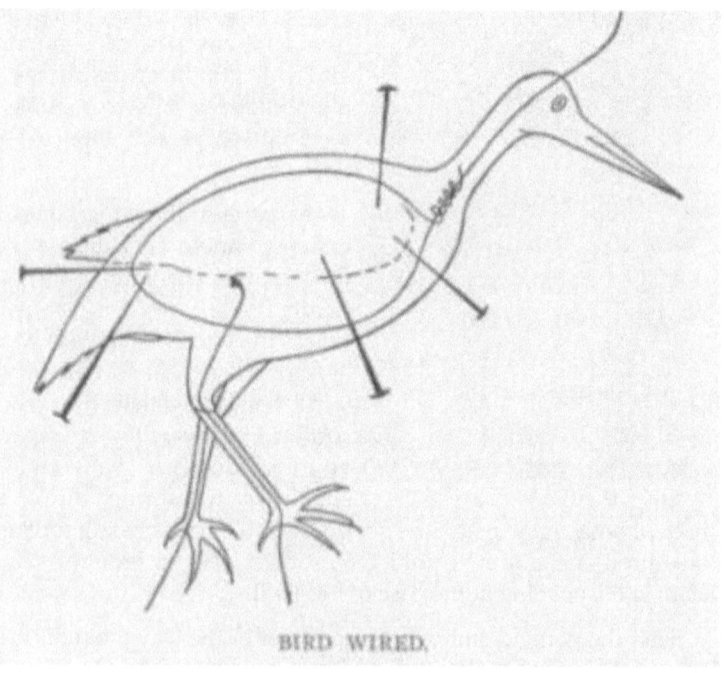

BIRD WIRED.

Work this loop back through the body pulling back through the leg and clinch the points over into the body again. If the body was firmly made as it should be, this fixes the legs permanently to it, yet they may be bent readily at the joints to suit various attitudes.

Adjust the skin now and fill out at shoulders, thighs, and base of tail with some chopped tow. The breast also may need some filling. Sew up the skin beginning at the breast and finishing at the base of tail, lacing it together with the ball cover stitch.

The pose of the finished specimen you have probably decided on before now so it only remains to put it in the desired attitude. This sounds easy, like sculpture, "just knock off what stone you don't need and there is the statue," but it may try your soul at length to obtain the desired effect. Good pictures are a great help here, as of course a living duplicate would be if you had it.

Before setting it on its feet, coax the wings into place as you hold the bird in your hand and pin them to the body through the elbow and outer joint of the wing, using several pins 2 to 4 in. long or pieces of sharpened wire the same size. This will hold the wings out of the way and they may be changed and fitted perfectly later.

Mounted birds are usually put on temporary stands of rough boards or limbs and when fully dried out transferred to a permanent mount which can be prepared in the meantime of the exact size and variety wished for. On these temporary stands the leg wires are only twisted together so they can be easily removed. Place the feet in natural positions of standing, walking or running and arrange the toes correctly. They had best have some pins driven in beside the toes to secure them till dry, as badly shaped feet will spoil the effect of an otherwise fine piece of work, indicating a careless workman.

If on a bough or stump the feet should grasp it as if the bird really means to stay on it. Two or three wires like those used on the wings hold the tail in place by being driven through the base of it into the body for half their length.

Fix the head looking down rather than up and to one side rather than straight ahead. If you have the proper glass eyes at hand they can be set now, if not, later will do but the lids are relaxed just now to receive them. Fill the back of the sockets with tow or cotton and with a little spoon-shaped modeling tool give this and the inner surface of the lids a good coating of soft clay. The eyes, cut from the wire stem on which most of them come, are pressed into this and the skin worked into place with the point of a big needle or a small awl.

Now give the plumage a general going over, re-pin the wings if necessary, and wind down any obstreperous feathers with thread. A number of pins or wires thrust in the middle of back and breast will help this operation.

Starting at the head wind back to the tail, lacing the thread from pin to pin, not binding tightly with any one thread but producing a smooth surface by holding it down at a multiplicity of points. There are a number of so-called systems for winding birds but the same taxidermist seldom winds two alike as the needs of the case are sure to differ. To spread the tails of small birds, spread the feathers as desired and pin them between two strips of light cardboard. When dry they will retain their position. If all arranged properly set the bird away to dry; two weeks will be sufficient for this.

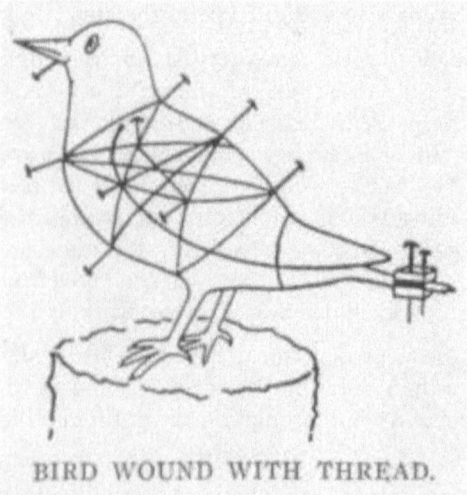

BIRD WOUND WITH THREAD.

Any colors on bill or feet and legs should be renewed with oil colors as they fade almost entirely; if of a dark or neutral color originally, a coating of transparent varnish will do. There is a variety of beetle which delights in dining on such hard parts of mounted birds if not protected by paint or varnish.

Place the bird on its final mount and fasten the leg wires in grooves cut in the under side of same so they are flush with surface. The ends may be turned over and driven in again or held fast by small staples. If on the under side of a limb or branch a pinch of moss or lichen glued on will cover this fastening.

Cut away the binding threads and with cutting pliers cut off all projecting pins and wires, leaving what part of them is covered by the feathers. Brush any clay from the eyes and if the lids have shrunken away from them as they do usually, coat a bit of black thread with glue and with a setting needle place between the lids and glass eyes.

THE POSE OR POSITION OF CERTAIN BIRDS

If the eyes were not on hand when first mounted the lids should have been relaxed and the eyes set as soon as procured.

Small birds may be mounted in fluttering attitudes by pinning the wings

FLUTTERING POSITION OF SMALL BIRDS.

with a bunch of cotton beneath them, but if an attitude with wings fully spread is called for they must be treated in a manner similar to the legs, inserting the wire under the skin near the tip of the wing and running it along the under side of the wing bones to the body through which it is thrust and clinched as the leg wires were.

Birds mounted with spread wings cannot be so readily wound to smooth the plumage and require to be braced with strips of stiff paper and sometimes extra wires, which are removed when dry.

NATURAL STANDING POSITION OF SMALL BIRD ON LIMB.

CHAPTER X.

MOUNTING LARGE BIRDS.

The same principles employed in the manipulation of smaller species apply to this class with but a few variations. On account of their size and weight the artificial bodies need to be especially firm in order to afford a secure anchorage for the wing and leg wires. Also these supports should be fastened to the bones in several places with stout cord or small wire, as wrapping with the tow and thread used on small birds is hardly sufficient.

What I would class as being large birds are the larger hawks, owls, herons, eagles, geese, etc. The several varieties of the ostrich are known as colossal birds and are skinned and preserved much as the large quadrupeds by mounting the prepared skin on a manikin, built in the pose of the finished specimen and supported by heavy iron rods.

BIRD OF PREY—LIMB POSITION.

In mounting eagles or similar birds with wings spread, which seems to be a popular attitude, use the largest wires possible as anything less than that will, on account of their size and wide extent, tend to a drooping, back-boneless appearance entirely out of keeping.

It goes without saying that large birds do not require the delicate handling of small ones, but by way of compensation considerable force is needed.

The combs and wattles of domestic and wild fowls cannot by any common process be prevented from shriveling and discoloring while drying, but when dry they may be restored by careful modeling in colored wax. This is applied warm with a brush and given its final finish with hot metal modeling tools. For museum work and other high grade work such heads are cast entire in wax in such a way that all feathers and hair are attached in their precise places.

SPREADING TAIL OF LARGE BIRDS.

Run a small sharp wire through the quills on under side of tail to spread it.

Large water fowl are often mounted as flying, and suspended by a very fine wire. A sharpened wire with a ring turned in one end, thrust into the middle of the back and clinched in the body, forms a secure point of suspension.

As it is not usually practicable to case many specimens of large birds, give them an extra thorough poisoning and when entirely finished spray with either corrosive sublimate or arsenical solution.

In making bodies for large birds it is well to use excelsior for the main bulk of the body, merely covering the outside with a thin layer of tow. This is not only more economical but makes a lighter specimen than one filled with tow entirely. Excelsior or wood wool is to be had in varying degrees of fineness of upholstery dealers.

SPREAD EAGLE, WINGS BRACED
UP TO DRY.

In the case of a bird which has been wired and sewn up seeming to require further filling out, it can be accomplished in most cases by making an incision under each wing and introducing some flakes of tow with a wire stuffing tool. If the bird is mounted with closed wings this slit need not even be sewed up as the folded wing covers it completely.

CHAPTER XI.

TANNING, CLEANING AND POISONING SKINS.

I have used the following method for some years successfully on skins up to and including the deer in size. Most larger skins need thinning with a special tool, though an experienced hand can manage to thin a heavy hide with a common draw knife.

An empty lard tub, a half barrel or a large earthenware jar to hold the tan liquor, a fleshing knife and a fleshing beam are necessary to begin with at least. Any smith can make a knife of an old, large file or rasp by working both sides to a blunt edge and drawing the upper end out in a tang for another handle. A piece of old scythe blade with cloth wrapped around the ends will do, or a dull draw knife, either. One blade filed into fine teeth will be useful in removing the inner or muscular skin.

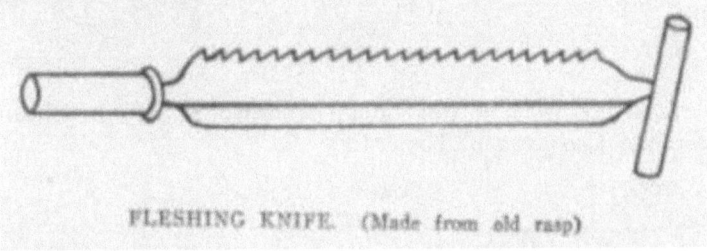

FLESHING KNIFE. (Made from old rasp)

A slab or plank 6 or 7 feet long, with one end tapered and half rounded, on 2 or 4 legs of such length as to bring the end against the workman's chest, makes a beam.

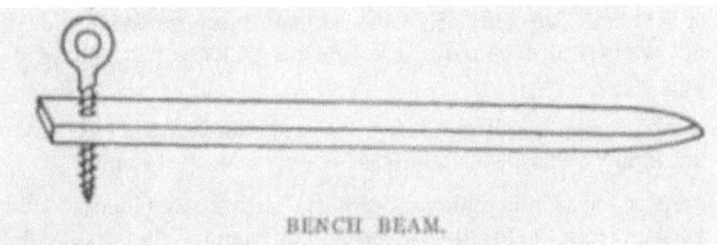

BENCH BEAM.

A short piece of plank rounded off and bolted to the top of the table or work bench will do for small skins.

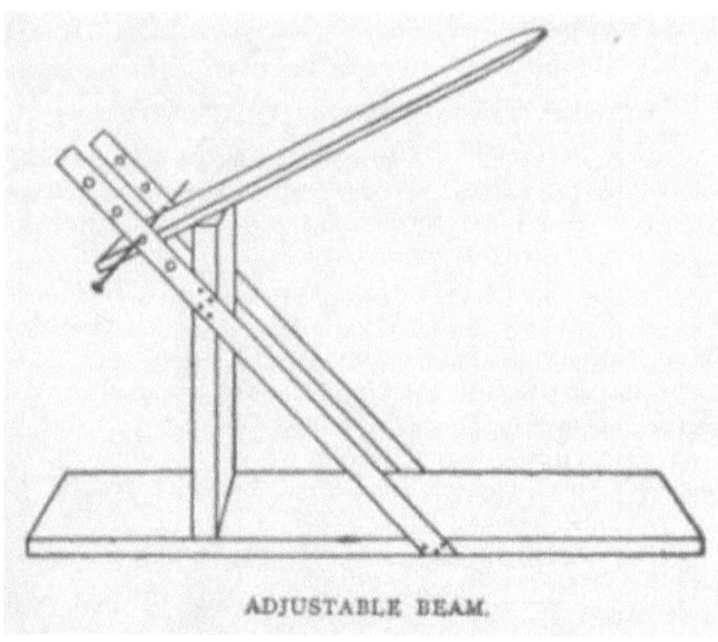

ADJUSTABLE BEAM.

Make up the quantity of tanning liquor you wish by the following formula, also given in Chapter III:

Water   1 gallon
Salt   1 quart

Bring to the boiling point to readily dissolve the salt, and add

Sulphuric acid   1 fluid ounce

Allow it to cool before putting in the skins. *Handle the undiluted acid with care.* Use common coarse salt and the commercial acid. Skins not wanted at once may be left in this pickle for months without injury.

About a gallon of pickle to a fox size skin is the correct amount, but it may be used several times before losing its strength.

After the skin is removed from the carcass any blood and dirt is washed from the fur, the flesh side well salted, rolled up and left 12 to 48 hours. Then thoroughly beam or scrape down the inside of the skin, removing all flesh, fat and muscles. Skins already dry may be placed either in clear water or tan liquor until they soften up. It takes longer to soften in the tan, but if put in water it must be watched or the hair will start, especially in warm weather.

A very hard, dry skin must be soaked and scraped alternately until thoroughly relaxed. When well softened, treat same as a fresh skin. If very fat or greasy, soak the skin in benzine an hour, wring out well, hang up till the benzine is about evaporated, then place in the tan. If not very fat the skin need not be put in the benzine first, but go in the tan at once after being scraped. I use common stove gasoline for benzine; it is as good for the purpose of cleaning and deodorizing, and cheaper. It must never be used in the vicinity of fire or a lighted lamp, as its evaporation produces a very explosive gas. For this reason I do the cleaning and beating of furs out of doors if practicable. Gasoline wrung out of skins may be saved and, after allowing to settle, used again and again. It will not do for the final cleaning of white furs, but for removing grease before tanning, it is as good as any. Stir the skins about in the tan frequently that all parts may come in contact with the solution.

Light skins like the fox will tan in 24 to 48 hours; heavier ones in proportional time. When on pulling or stretching the flesh side, it whitens, it is tanned. On taking from the tan, rinse the skin well in lukewarm water containing a handful of washing soda to the bucketful. Wring out with the hands and soak again in benzine for half an hour. Wring out of this and clean the fur at once as follows:

Fill a shallow box part full of clean sand or corn meal which has been heated in a pot or pan over the fire or in an oven. Work the skin about in this until the fur is filled with the meal; then shake or beat it out. Repeat the working in meal and beating until the fur is clean and bright. Never put a damp skin into the meal without the gasoline bath first, or you will have the time of your life getting the meal out. Regular fur dressers use very fine saw dust, but meal is to be had anywhere. Plaster of paris will do, but it is most too fine. This treatment with gasoline or benzine removes the grease and animal odors so apt to cling to Indian or home-dressed skins. After cleaning, allow the skin to partly dry in the air and shade; then soften it by stretching, pulling and rubbing in every part. There is no way to accomplish this without work. A pad made of old bags, pieces of blanket, etc., put on the beam, the skin placed on that and stretched in all directions with the blunt edge of the fleshing knife is as good as any way. Keep up the stretching until the skin is quite dry. If any part should dry out too fast for the operator (that is, before it gets stretched) it can be dampened with water on the flesh side and then treated like the rest. If it is wanted extra soft the skin may be thinned down with sandpaper. If the dressed skin is wanted to lie flat as for a rug, it can be moistened on the flesh side; then stretched out and tacked fur side up on a board, the table top or the floor until dry. If this should cause it to harden or stiffen too much break it again without stretching too much.

Another very good tanning solution is the following:

Salt	1 quart
Oxalic acid (pulverized)	4 ounces
Water	2 gallons

Dissolve well and immerse the skins, treating them as already directed, rinsing in clear water only. It is also best to allow a little more time for tanning in this solution.

While on the subject of dressing skins a few words in regard to cleaning furs will be in order. White furs especially that have become soiled and matted from use need cleaning frequently and are brought to the furrier or taxidermist for that purpose.

A good washing in gasoline will usually remove the dirt, then dry out as in dressing furs. Furriers often use powdered magnesia for this purpose but almost any finely divided white powder will do about as well. A long siege of beating, shaking and brushing will be necessary to get the drying powder all out of the fur so it will not sift out on the garments when wearing.

If the piece to be cleaned is large (like a coat or cloak) the lining should be removed before cleaning. Neck pieces and small furs are cleaned with linings, wadding, etc., intact. If the fur is so matted that beating does not fluff it out, it must be combed, using a metal fur comb to break up the tangles.

Charges for this work are based on the time used, though at least 75c. or $1.00 each is charged for the small pieces.

As chemicals exposed to the changes of atmosphere are likely to lose strength in time, the owners of mounted heads generally take the precaution of having them poisoned against moth at varying intervals.

Personally I think once in three years is sufficient but some prefer to be on the safe side where valuable heads are concerned and have them treated yearly.

Large heads like buffalo, moose, and elk may be poisoned as they hang, thus obviating any subsequent handling which would be to the detriment of both specimen and operator.

Heads the size of deer and smaller are readily removed and replaced.

First dust the head well and comb the hair or pelt to detect the presence of moth. If loose hairs reveal this the head should be removed to the open air, unscrewed from the shield and saturated with gasoline, which will kill both the moth and its eggs, after which poison against a repetition of the offense.

Pin an old cloth or newspapers about the neck to protect the shield and wall and spray the entire skin of the head with the diluted arsenical solution as recommended in Chapter III. Seedsmen sell a sprayer for use on plants which is about the most convenient size, though the larger size used in the vegetable garden or even a toilet atomizer will distribute the solution.

After it has dried wipe off the eyes and nose with a damp cloth and handle as little as possible.

The common tariff on such work is for treating a single head not less than $1.00. Two to four at one place, 75c. each, and over four, 50c. This for poisoning only. Extra charges for killing moth. Such work should be done in spring or early summer in the Central States in order to be effective.

CHAPTER XII.

MAKING ANIMAL FUR RUGS.

Probably the first use (after clothing) made of skins was as rugs or coverings for the ground or couches, and in this shape they are still to be found in our most elegantly furnished homes. One of the few survivals of primitive tastes.

The skins of some few animals such as Polar and Grizzly Bears, Tiger, Jaguar, Lion, Puma, Leopards and Ocelots are used for little else, though some of the spotted cats are used for eccentric looking coats and fur sets. Other smaller skins such as wolf, fox, 'coon, wild cat, etc. are much in favor as rugs as well as for garment furs.

In skinning an animal for use as a rug it is as well to skin and stretch it open, cut under side of body from chin to the end of tail and from each foot down to the central line. A large animal like bear or leopard looks well with the paws preserved and they should be skinned down to the last joint, leaving the claws attached to the skin. Smaller skins may have the paws preserved, though the effect is hardly worth the trouble and the smaller paws are easily crushed on the floor by a chance step.

After skinning, using care to detach it from the head without mutilating the ears, eyes and lips, stretch flat on an inside wall, door, or table top. Stretch evenly with tacks or small nails close together to avoid drawing out in points and of the approximate shape of the finished rug. That is, with the front feet well forward and hind feet pointing back, not spread as wide as possible.

If you are intending to dress the skin it may be begun at once after skinning, as per the chapter on tanning, etc., or after fleshing it may be put in the pickle jar against a leisure day. Otherwise stretch and dry for transportation or to send to the tanner.

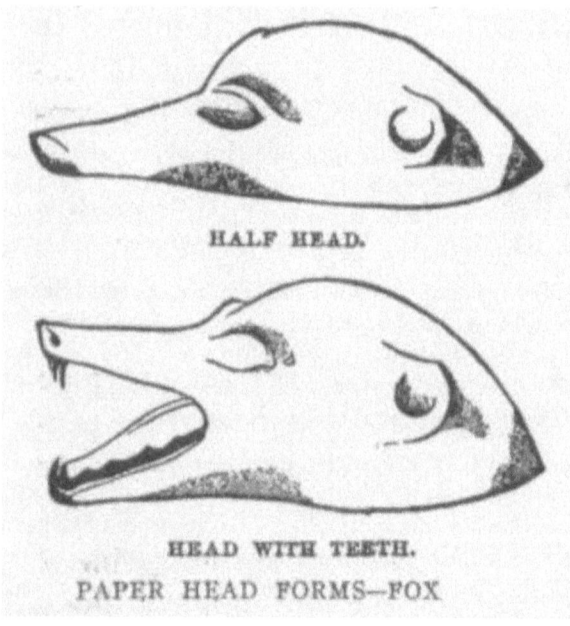

HALF HEAD.

HEAD WITH TEETH.

PAPER HEAD FORMS—FOX

As regards the mounting of heads for rugs, they may be done in three styles, called half or mask head, full head closed mouth, and full head open mouth. The first, as the name indicates, consists of the skin of the upper part of the head without that of the lower jaw mounted over an artificial form or "skull." The closed mouth (See dog) has the lower jaw mounted in addition, but without any teeth used, and the open mouth mounting requires a set of suitable teeth with the interior of the mouth, tongue and lips fully modeled and finished either with colored wax or by painting.

These artificial head forms or skulls both with and without teeth and masks, are to be had in all varieties and several sizes each of dealers in taxidermists' supplies so cheaply that I would advise the novice to procure them if possible. In many cases it is necessary for the professional to make use of skulls with artificial teeth as the natural skulls are often thrown away by the collector. In the case of any large skin intended for a rug the roughly cleaned skull should accompany same. In ordering from dealers it is only necessary to give name of animal and the measure of skin from center of nose to inner corner of eye, and outer corner of eye to ear.

The beginner would do well to try mounting a rug with half head first and the more difficult open mouth later. A very fair mask form can be made by laying the skinned head down on a piece of thin board and marking around it with pencil, then cutting

DOG—CLOSED MOUTH.

out to the outline. With a bunch of fine excelsior or coarse tow and a spool of thread a half-head form can be roughly blocked out by winding, using the board as a base. Then with modelling clay and chopped tow the anatomy is perfected, pressing down here with the fingers, and building up elsewhere. With the skinned head to refer to as the form is modeled a good job can be done. However, if a number of skins of the same species are to be prepared it is best to make a mould in which unlimited paper forms may be cast. Particulars in this work are given in Chapter on Casting and Modelling.

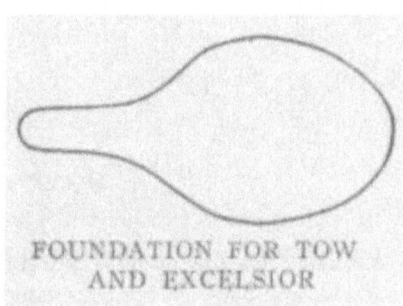

FOUNDATION FOR TOW AND EXCELSIOR

The next step in rug making after drying the pelt is to prepare the head for mounting. It is as well to do little or no thinning down of the head skin during the tanning and even if it has been shaved down the vicinity of the eyes, ears, nose and lips will need thinning with a small sharp knife, and stretching out with a skin scraper. Before beginning this process the head skin should be dampened on the pelt side with clear water (use pickle in warm weather). If the ears are not skinned before dressing they should be now, and turned inside out to the tips. A small screw driver with the edge blunted and rounded is a good tool for this work as it will not readily cut the thin skin of the ears.

Trim and scrape away any lumps of muscle, etc., and shave down the skin enough to be molded to the surface of the form when dampened. Do not, however, cut away the bunch of muscles on each side of the cheeks in which the whisker roots are embedded, or these distinguished ornaments will drop out. By criss-crossing these with cuts they are made as flexible as the rest of the skin. After the shaving process get a suitable needle and stout thread and sew up any cuts or tears that have been made. If proper care has been used there will be little of this to do, always remembering that a cut is not irreparable but always makes extra work. Bullet holes of large caliber destroy considerable skin and in order to close them it is best to cut them to a triangular shape and draw together by sewing up from the corners of the triangle, as per illustration.

Cut out from tough cardboard two ear forms a little longer at the base than the ear skin and small enough to slip inside them readily. Before going further give the inside of the head and neck skin a coat of preservative. Let this lay a few minutes to soak in and then after turning the ears right side out slip the cardboard ear forms into place. They should be coated first with liquid glue; work the skin over them with the fingers and fill around their bases with some cut tow and clay of about the consistency of soft putty. Now place the head skin on the form, get the eyes and nose in place and drive in a few pins down the center of the face; they will hold it from slipping while working further on it. If the form is a little too short for this particular skin build it out with clay and tow, if too long it can have a trifle cut off.

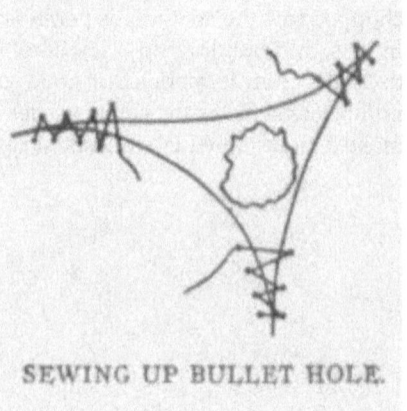

SEWING UP BULLET HOLE.

Fill the sockets of the eyes with clay, build out the cheeks and the sides of nose with clay and tow and draw the skin of the lips down where they belong. The glass eyes are to be cut from their wires and set, drawing the lids around them with an awl. When they seem properly placed drive pins at both inner and outer corners. The ears are attacked next and when arranged to suit, three or four pins driven in at their bases to hold them.

On a rug the ears should usually be laid back close to the head as by so doing the chance of their being broken off when finished and dry is lessened. Also a mounted rug head is usually intended to register rage or anger.

The upper surface of the face being attended to, turn it upside down on a folded bag or something in the nature of a cushion while we lace it across the form with a stout thread and needle. If a hollow paper form is used it should be filled with crumpled paper, excelsior, coarse tow or similar material. Do not use fur scraps for this as I have seen done or it will be a moth nest.

The whole inside of the skin may now be poisoned after slightly dampening, and then tacked out fur side up in the proper shape to dry. In order to make an animal skin lay flat to the floor it is necessary in most cases to cut out several V-shaped pieces. Behind the fore legs almost always and often in front of them, also and frequently in front of the hind legs are the places where these gores are removed. Consisting as they do of the thinly haired skin inside the legs their absence is not noticeable when neatly sewed up.

Take care in this final stretching of the rug skin to get it alike on both sides, or, as the artists say, bilaterally symmetrical. When tacked out, go back to the face and perfect it so it may dry just right. With a fine awl point draw the upper eyelids down a little, straighten the eye brows, lashes and whiskers, and mould the nostrils into shape, bracing them with damp clay; when dry it is easily removed. Now set it aside until fully dry before proceeding with the trimming and lining. One and a half or two inch wire brads are good to use in stretching skins, but 3d wire lath nails will do; the longer brads are more easily handled.

After removing the nails turn the skin on its back and draw a line from neck to tail with pencil or chalk. By measuring from points on this line we can trim off the legs and flanks of the rug evenly. If it is a small or medium size skin it will look best with an all felt lining. So by laying it flat on a piece of felt somewhat larger all round and marking around it at a distance of 3 inches we can cut out the lining. The edge of this is to be pinked. One end of our chopping block, usually of sycamore or oak, is kept for this function, and a few minutes work with pinking iron and hammer will border the lining with neat scallops.

A sufficient length of felt strips about 2 inches wide, should be cut to reach around the outside of the skin, also pinked on one edge. Allow generously for this as it will have to be gathered in rounding the feet and head. In the case of animals having a bushy tail or brush as the fox, wolf, etc., the tail is merely sewed up on the under side after poisoning and not lined or trimmed. Pumas, tigers and others with short furred tails are trimmed and lined like the rest of the rug. In lining large rugs a double trimming of felt is often used and a lining of strong canvas is used throughout, as when on the floor it is not visible, protects the skin as well, and costs somewhat less.

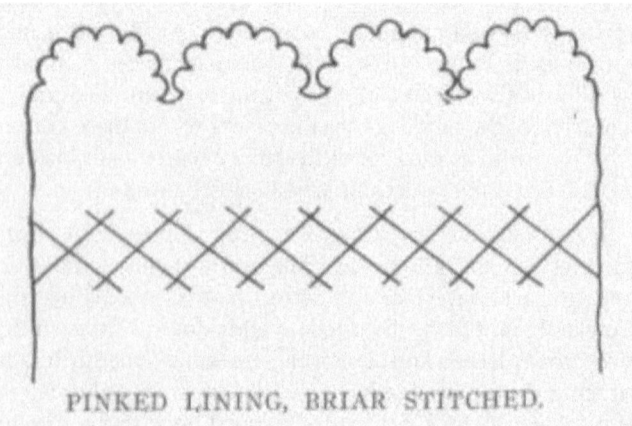

PINKED LINING, BRIAR STITCHED.

The trimming felt is sewed around the edge of the skin, passing the needle through from the back obliquely, resulting in a long stitch on the felt and a short one on the fur side. What few hairs are drawn down by this can be picked out later with a needle or awl.

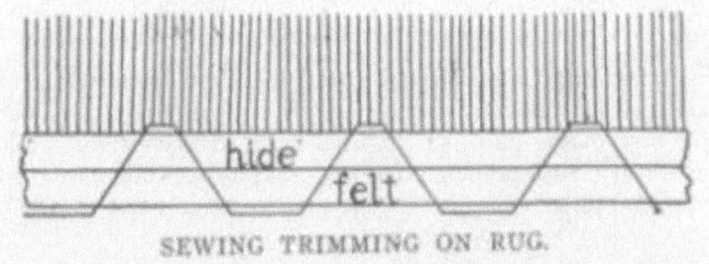

SEWING TRIMMING ON RUG.

Before sewing the lining on an interlining of cotton wadding should be cut out and basted in place with a few long stitches. Now place the skin fur side up, on the lining and adjust it so an equal margin shows on all sides and pin it in several places to prevent its slipping while sewing it fast. To do this turn it felt uppermost and sew around just at the edge of the skin, in the trimming felt, reversing the stitch previously used. This hides the short stitch outside and if drawn up evenly will hardly be noticed if a color of thread has been selected corresponding somewhat to that of the felt.

If an extra nice finish is desired the lining may be put on with a decorative briar or cat stitch with some bright colored silkatine.

Brushing away any clay from around the eyes and nose, giving the latter a touch of the proper color (black for the majority of animals). A coat of thin shellac to simulate the natural moist appearance and connecting the dried eyelids with the glass eyes with hot colored wax will about complete the rug. Waxing around the eyes is done with a small round artist's brush and adds to the finished appearance of a job.

In mounting a rug head with either full head, closed or open mouth, the beginner had best use a head form from the dealer for a few times at least. A little study of one of these will enable him to model an open mouth head, when a good set of teeth are supplied, and the ready made article not at hand. It requires considerable time and some natural ability to set the teeth and model the gums and tongue effectively.

A tongue modeled with clay and tow, covered with several layers of papier mache and when dry, coated with flesh colored wax is good enough for any rug, though museum mounting might require that the tongue be skinned and the skin used to cover the model.

Plaster, putty, papier mache and various plastic cement materials are used for modeling mouths, of which papier mache is probably the best; plaster paris is often used in an emergency but is brittle and heavy. For modeling use finely ground paper pulp mixed with glue and plaster or whiting. Only practice and experiment will determine just the precise mixture wanted.

A paper half head form may be the basis and to this wire the jaw bones with their sets of teeth. Clever work will reproduce the interior of the mouth, gums and tongue, and when perfectly dry they should be finished either with paint or colored wax.

The tongue should have its base and lower side coated with glue and have a brad driven through it into the material between the lower jaw bones. If the head of this brad is well set in, a drop or two of wax will cover it.

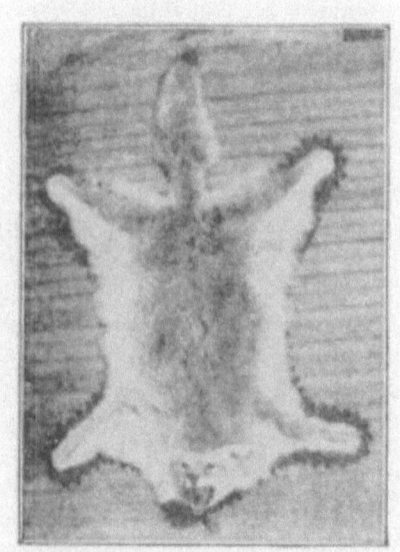

COYOTE RUG, OPEN MOUTH
(Author's Work.)

In preparing a skin for mounting an open mouth head the lips should be pared down and preserved as far as possible as they are to be filled out and attached to the form by pinning at their edges. Common toilet pins are used for this, driving them in part way and when the work is dry cutting them off close down to the surface. After this is done the lips may be waxed thus joining them to the form completely.

Never fear to use plenty of pins in head mounting. In some places they may be driven to the head and left covered by the fur, in other places where there is little or no fur, cut them close and drive down flush.

Of course greater liberties may be taken with a rug skin than one mounted entire for exhibition, still a competent artist can put a great amount of expression in even a rug head. The close student of animal anatomy can produce an appalling snarl of anger on the heads of the larger carnivora or change the same to a sleepy yawn or grin in a few minutes' manipulation.

The professional is often called in to repair damaged rugs and especially those with open mouths. Here the operator must use his own judgment as no two seem to demand the same treatment. Missing teeth may have to be supplied and carved from bone, celluloid or antlers. The tips of broken deer antlers make very good canine teeth and blocks of celluloid which are much easier to shape than bone, are sold by supply dealers.

I have dwelt at some length on rug making as it is a branch of taxidermy which seems to be always in more or less demand with the public. Also it forms an easy entrance to the more complicated mounting of complete animals and much of the work is identical with the process of preserving heads for wall decoration.

CHAPTER XIII.

FUR ROBES AND HOW TO MAKE THEM.

While not usually classed as taxidermy the making and repairing of robes will bring in many a dollar to the worker in the middle and northern states. A stitch in time (on a robe) often saves more than the proverbial nine, and the better the quality the more anxious the owner to have it put in good order.

The late lamented bison furnished the robe par excellence, few of which pass through the hands of the taxidermist nowadays. Their place has, in some degree, been taken by the Galloway and other cattle hides, which also make a practically one piece robe of good weight leather. These are too heavy for economical dressing by hand, but the regular tanning concerns will dress them soft, pliable, and clean for a very reasonable price.

The regular robe makers do much of their work with the heavy overstitch sewing machines, but it can be done as well or better by hand at the expense of more time. Many of the smaller skins, as coyote, raccoon, fox, opossum, and wild cat make up as handsome carriage robes and sell at remunerative prices.

Skins of an inferior lustre or that are mutilated are often used. For instance, the skin of the head may be mounted separately and not interfere with using the balance in a robe. For use in a robe skins should be taken off open and stretched in a rectangular shape as near as possible.

After tanning, sew up all cuts and holes in the skins, dampen the flesh side with clear water and tack out fur side down on the floor, table top, or better still on light boards cleated together which may be set on edge against the wall out of the way. In all sewing on rugs and robes be sure and use a substantial thread well drawn up, fine stitches are not essential but good material is, as such things come in for a deal of rough use unlike mounted specimens which are, or should be seldom handled. Glovers triangular needles and gilling or carpet thread of suitable sizes are the necessary tools.

Skins of approximately the same size should be used in making up a robe or the effect will be bad. After stretching and drying, cut them to rectangular shape, taking care to get the darker line down the back in the center of each. A good way is to cut a piece of cardboard to the required size and mark around it. Gaps in front of and behind the legs may be filled by sewing in small pieces rather than cut down the skins too much. The drawing shows coon skin marked to cut for robe. The skin is poorly stretched yet there are many even worse, altho trappers are learning to handle the skins in better shape.

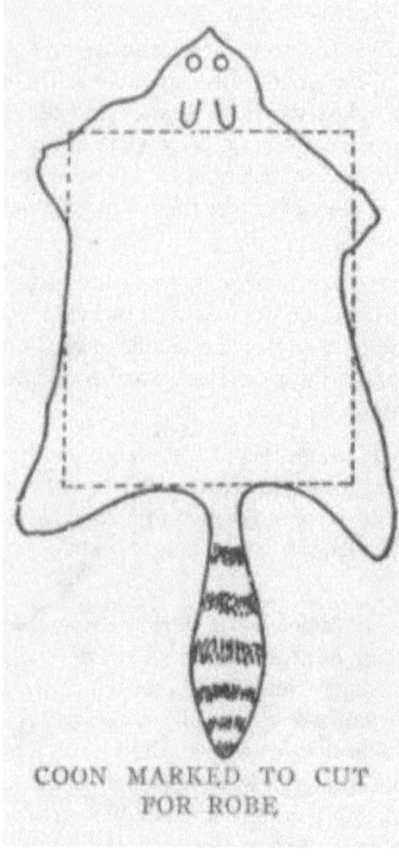

COON MARKED TO CUT FOR ROBE

After the necessary number of skins are cut out they are sewed together in rows and the rows in turn sewed to each other like a patchwork quilt, taking care to have the fur all run the same way. The robe should now be dampened again and stretched and tacked to its full extent to remove any wrinkles and flatten the seams. This sewing is all done from the back of the robe using an even overhand stitch. Just before the final stretching it is well to apply arsenical solution to the damp skins.

On drying it is ready for the trimming and lining. Sometimes it is necessary to turn over the upper edge of the skins and baste it down or it may show the raw edge of the hides on account of the fur laying all one way.

Strips of felt 3 in. wide, pinked on one edge are used for trimming, and may be had from the dealers, ready to use, or we can buy the felt by the yard and easily pink it in the work room. Any one doing much robe work would find a pinking machine (price about $5.00) a good investment, but the small hand iron does just as good work.

A double border of contrasting colors may be used, the upper strip of which should be inch narrower. The border is sewed on from the back with heavy thread, using the same stitch as that for the lining rugs. An interlining of cotton wadding is basted in place before the lining is sewed on. Plush or beaver cloth is to be had in 54 and 60 inch widths and a variety of colors, of which the darker greens, browns, and blacks are to be preferred.

STRONG HIDE (CATTLE) LAP ROBE.

Fur robes are usually 48☐0; 54☐6; 60☐0; or 60☐4 inches in size, so linings of the above widths cut without waste.

Single cattle or horse hides may be left with the edges merely straightened or they may be cut and pieced to regular robe shape. The bushy tails of such animals as wolf or fox are sometimes used along one side or both ends as a fringe.

The number of skins required to make a robe are as follows:

Raccoon	12 to	20
Coyote, wolf or dog	6 to	10
Wild cat	12 to	16
Woodchuck or opossum	20 to	30
Goat	4 to	8

COYOTE LAP ROBE, 8 SKINS.
Note tails in center.

Baby carriage robes of angora or lamb skins are lined with quilted satin and trimmed with felt of some light shade. They usually have either an opening for the head and shoulders or a pocket for the feet.

The natives of Patagonia make up many robes of the guanaco and vicuna, dressing the skins and sewing them together with sinew. Their dressing is faulty as the skins are apt to stiffen and crack and the sinew hardens with time until it becomes like wire, though the stitching is wonderfully even. They have, however, worked out a scheme of joining the skins in a way to eliminate waste, that is far ahead of civilized fur workers. A row of skins are joined head to tail and the next row headed the opposite way will fit in perfectly, the legs being left on the skins. The sketch with this will explain better than any description. The guanaco pelt being of a woolly nature makes it unnecessary to run it all the same way and the entire skins are utilized in spite of their ungainly shape, the flaps and tabs trimmed off filling the indentations around the outer edge of the robe. They make an excellent camp blanket as light and warm as the malodorous, hairy rabbit skin robe of Hudsons Bay, and no Patagonian ranch house bed is complete without its guanaco coverlet.

PATAGONIAN ROBE OF GUANACO SKINS.

You will likely be called on to repair robes much oftener than to make them and such work is nearly all profit, as it generally consists in sewing up rips and tears in the skins. Never attempt to do this from the front or fur side as it can only be done right from the back. To do this at least one side of the lining will have to be ripped and the robe turned, turning it back and resewing it on completion. Linings are turned under at the edges all around.

Worn and soiled linings and trimming often need replacing with new material and it is sometimes necessary to purchase an unlined goat "plate" to repair robes of that common variety. Worn robes can be cut down in size if no similar material is to be had for repairs.

CHAPTER XIV.

MOUNTING ENTIRE SMALL FUR ANIMALS.

In Chapters VI and VIII directions are given for skinning and preparing this class of animals for mounting, so with the skin properly cleaned and poisoned before us the next thing is to cut the wires for a supporting frame. These are six in number usually, body wire, tail wire and one for each leg. The body wire is about one-half longer than from nose to base of tail; tail wire the length of the tail bone and half the body, and each leg wire twice the length of the leg.

I have spoken of using a muskrat for an initial attempt as it is of a convenient size to handle and the length of its fur will hide small defects in the anatomy. Most books of instruction select a squirrel for the beginner's victim. It is true it is not as difficult as a hairless Mexican terrier but it is apt to discourage the learner. An opossum will do very well or any long haired animal of about that size.

We will first reconstruct a hind leg and if it is a fresh specimen being mounted without a bath in the pickle we can have the opposite leg in the flesh to guide, as to proper proportions. The wire is passed through the cut in the bottom of the foot and along the back of the leg bones where it is secured in about three places by tying with small cord. The end is left projecting three inches beyond the end of the upper leg bone.

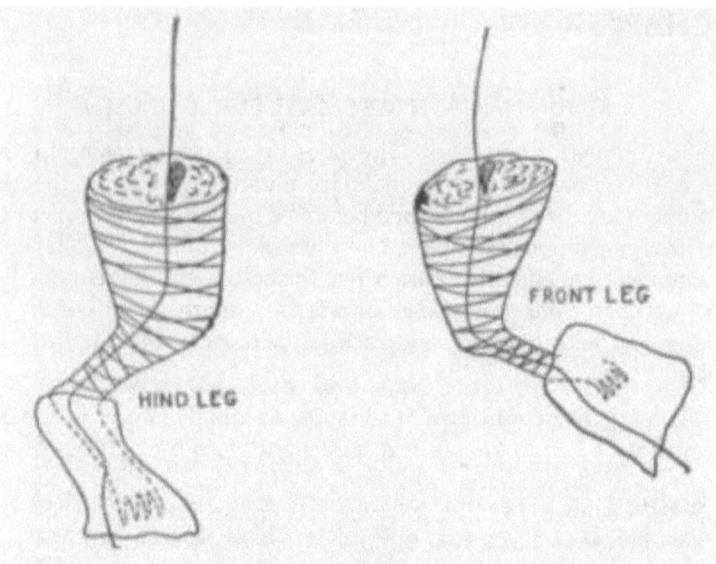

HIND AND FRONT LEG OF MUSKRAT WRAPPED READY TO CLAY AND RETURN TO SKIN.

The muscles and flesh are replaced by winding on tow with stout thread until the approximate size is reached, when the surface is given a thin coating of clay and the leg is drawn back into its skin. The fore legs are in the same manner built to the joint with the shoulder blade.

After sharpening the tail wire to a point it is wound with fine tow and thread and coated with clay until it duplicates the bone and flesh removed. This is slipped into the tail sheath with the unwound end projecting into the body and the slit along the lower side of tail sewed up.

After making a ring about the diameter of a .22 shell on one end of the body wire place it on your sketch where the hip joint was marked, letting the wire run lengthwise of the body. Another ring similar is made at the shoulder. These form the points of attachment for the legs.

The skull, cleaned of flesh and poisoned, should have the muscles replaced with tow and the whole coated with clay. Force a piece of cork into the opening at the back of the skull. Sharpen the end of body wire and force it through the cork and out one of the nostrils. The skull is pushed back along the wire until it reaches the proper distance from the shoulder ring, when all but an inch or so of the projecting wire is cut off.

Insert the skull through the body opening and work it up the neck into its place in the head skin, letting

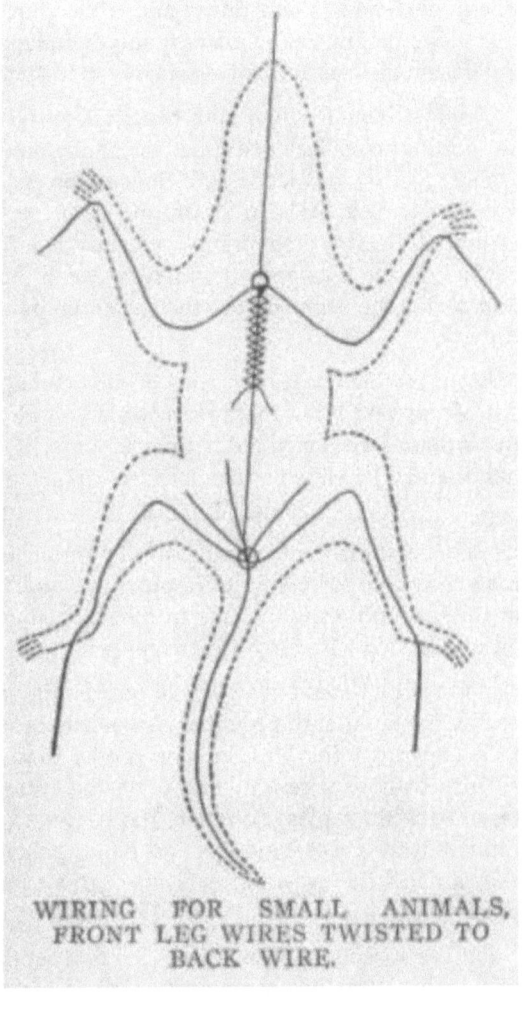

WIRING FOR SMALL ANIMALS, FRONT LEG WIRES TWISTED TO BACK WIRE.

the end of wire go through the nostril of skin also. This will hold the nose in place. Adjust the eyes and ears also.

Now pass the ends of leg wire through the rings from their opposite sides. The tail wire is passed through the rear ring and twisted around the body wire a few times. The ends of leg wires projecting through the ring cross, so twist them together a turn or two with pliers, next bring them down and under the body wire, twisting them together, first one side of it and then the other. This treatment will fasten the legs and tail also firmly to the body wire.

Thus we have the skin with head, tail and legs filled out and the supporting wires fastened together. The remainder of filling, usually chopped tow, is placed with fingers, forceps, and stuffers. First a good layer next the skin all around, then part this and fill in the center a little at a time, first at one end, then the other. Put a good cushion at the hips and shoulders between the wires and the skin and also at the back. Fill out the neck well but do not stretch it unduly.

Begin sewing up at both ends of the opening cut, drawing a few stitches up and tying the thread while you fill a little more. Model the animal into shape from time to time by pressure with your hands and when filled out and sewed up tie the threads together.

Our animal is now lying on its back with head, tail and limbs extended; bend them into some natural position and set it on its feet. It may be well to force a little chopped tow and clay into the bottoms of the feet and draw the cuts in them together with a few stitches around the leg wire projecting from them.

A temporary stand of a piece of board supported on cleats at each end is prepared and a number of holes bored for the leg wires. A little experimenting will find the proper place for these when the surplus ends of wire are bent along the bottom of the board and fastened with staples. Complete the head and face modeling now, filling out the cheeks and lips and pinning them in place. Work the skin around the eyes and ears into proper place and fasten with pins.

Fill eye sockets with sufficient clay and set the eyes, drawing the lids down where they belong.

Any deficiencies at the back of the head can be filled through the ears. In the case of a muskrat the ears are so insignificant that they only need a little clay and tow forced into them to hold them in place. More prominent ears as those of fox, 'coon, or wild cat must be filled with a pasteboard form, cut the proper size and shape, coated with liquid glue and inserted from the inside before returning the skull to the skin. The ears of all animals should be pocketed when skinned, that is turned inside out to their tips to admit preservatives and later some filling material which will retain their shape when dry.

Do not skin out and throw away the ear cartilage but leave it adhering to the skin of the inner side of the ear. Without it this skin is very frail and brittle and thorough pickling will prevent shrinkage and distortion of the ear.

OPOSSUM MOUNTED IN WALKING POSITION.

Before leaving the head push it slightly towards the body on the wire and cut same close to end of nose. Pull head back to place, the wire disappears up the nose about inch, then you can shape the nostrils and fill so they will not shrivel up in drying and look as though their owner had been a mouth breather.

If the general pose and appearance seem correct finish up by placing the feet and toes correctly. Nothing gives a mounted animal a more trampish, disreputable appearance than slouchy, run over feet with toes that don't seem to be on the job. Lastly comb the fur out and fluff it up before setting away to dry.

Animals up to the coyote in size are usually mounted by similar methods to the preceding. Sometimes a piece of board is substituted for the body wire, especially in the larger specimens, the wires to which are too heavy to clinch readily. The skull is on a separate neck wire and all wires are fastened to the back board by passing through holes and then stapling.

Of course it is possible to mount small specimens by the same methods most large ones are, by drawing the skin over a hard filling, in fact a statuette, which must be made to fit the skin. This method in the case of small animals requires so much time that it is impossible in ordinary commercial work.

CAT SITTING AND WATCHING

Strive to put your mounted animals in easy natural poses unless you are making a grotesque, in which case go the length.

Clean the eyes and teeth with a brush when dry, and beat the fur to make it stand out. Fasten securely on whatever form of mounting you have decided on, countersinking the wires on the under side. Accessories, as a piece of food in the mouth or paws, are added now if they have been prepared for.

A slip with record of the specimen written on it and pasted to the under side of stand will usually be appreciated. If the mouth is wanted open it should be braced in that position, the lips, etc., held in place by clay. When it is dry this can be dug out with awls and modelling tools and the tongue, gums, and inside the mouth modelled in mache or some plaster composition. The tongue may be modelled in connection with the lower part of the mouth or made separately and fastened in place with a brad and some glue.

Colored wax, pink for the inside and black for the lips, applied hot with a little brush in several coats finish the open mouth. A little black wax will join the eyelids to the glass eyes if they have shrunk away and the inside of the nostrils should be coated with a little pink. Bare skin on the end of the nose should be varnished.

CHAPTER XV.

MOUNTING LARGE ANIMALS ENTIRE.

Though at one time nearly all animals were mounted by the soft body or stuffing method as described in the previous chapter, very few of the larger ones are so treated now. An adequate frame is built in a body of the proper size and proportions, the surface of which reproduces those muscles lying next the skin. The skin, well pared down and poisoned, is sewed, pinned and glued to this surface.

In the small specimen clay was used next the skin in places to perfect the modelling, but such amounts would be required for a large animal as to affect the durability of the skin. Clay and plaster being in a dry state very absorbent, will eventually rob of all oily matter any skin in contact with them. Such skins will crack, split and finally disintegrate as thoroughly as those having an excess of fat adhering to them.

To prevent this a layer of some glue composition or paper is used just beneath the skin. As an example in this mode of mounting a black bear would answer nicely. If the leg bones are attached to the skin they may be unjointed at the toes and laid aside while the skin is well shaved down on its entire inner surface. A thoroughly flexible skin is entirely at the command of the taxidermist, one stiff or hard cannot be placed or kept in place at will.

After beaming, splitting the lips and nose cartilage, pocketing the ears and sewing up cuts and tears, the skin is dropped in the pickle. An outline sketch is made with chalk on the shop floor and on this the bones of the legs are arranged. A stiff wire bent along the back of each set of leg bones will guide us in bending the iron rods used as supports. These should be from 5/16 to inch in diameter, threaded and fitted with two nuts at the lower end and eighteen inches or so longer than the leg bones themselves.

Of this extra length, enough is allowed below the feet to fasten to the pedestal, the balance is bent in a right angle from the end of the upper leg bone. At the distance of the hip joint from the central line of the body it is bent again parallel with the back board; for a hind leg. The front leg rods are bent in the same way at the joining of the shoulder blade with the humerus or upper bone of the front leg. You will readily see the desirability of preserving at least one set each of the hind and front leg bones. In such case the missing bones can be roughly blocked out of wood to the proper dimensions, while if none are saved you will have to do the same depending on the skin for measurements.

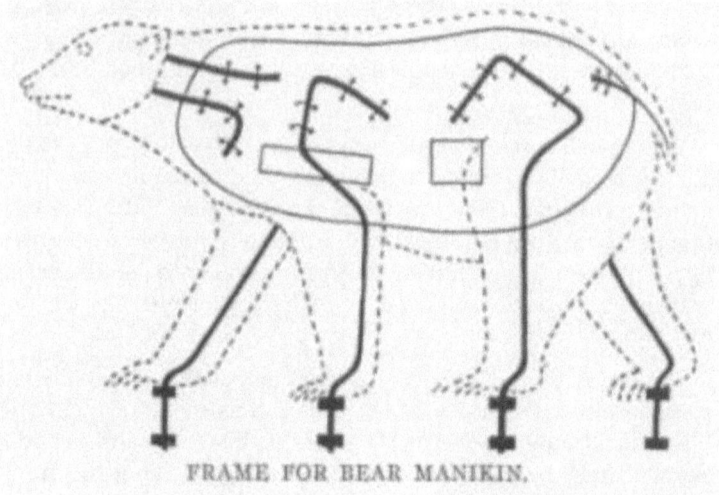

FRAME FOR BEAR MANIKIN.

The end of the rods lying along the back should be bent again in a V shape to prevent their turning when fastened to it. The location of the hip and shoulder joints are marked on one side of the back board, the rods for that side laid in place and fastened by drilling holes each side of them and passing loops of stout wire through and twisting them tightly on the other side with heavy pliers. The rods for the other side are fastened in the same manner, in fact they may be fastened with the same wires, but it will be stronger if the fastenings are separate. The leg bones are bound fast to the rods with wire or twine.

Holding the back board in the vise by the middle the leg rods with bone attached are adjusted to the position of the finished specimen. The threaded ends which project below the feet are bent straight down.

A rough pedestal of boards on 2" cleats at each end, is made, the frame placed on it and marking where the rods will enter, bore suitable holes to receive them. One nut is turned up each rod a short distance and after inserting in the holes in the pedestal the others are screwed up tightly from below.

Our frame now stands alone and rigid and should be viewed from all sides to correct any errors. It should not be too high, front or rear, and also having the back-board perfectly vertical or plumb. Insert two pieces of rod in the opening at the back of the skull and fasten them there by mixing enough plaster of paris and water to fill the cavity, to the consistency of molasses and pouring it in around them.

The ends of the rods should be bent or roughened to prevent them slipping out after the plaster has set. A surplus of plaster can be placed around the articulation of the jaws, at the same time holding them in place. These neck rods are to run beside and be fastened to the back-board as the legs were.

Let one remain straight and fasten it loosely so it may be drawn in and out the loops until the proper length of neck is formed, then tighten them and fasten the other rod also. Before fastening these try the skin over the frame, making sure it will cover in all directions. A tail wire stapled to the top of the back-board completes the frame.

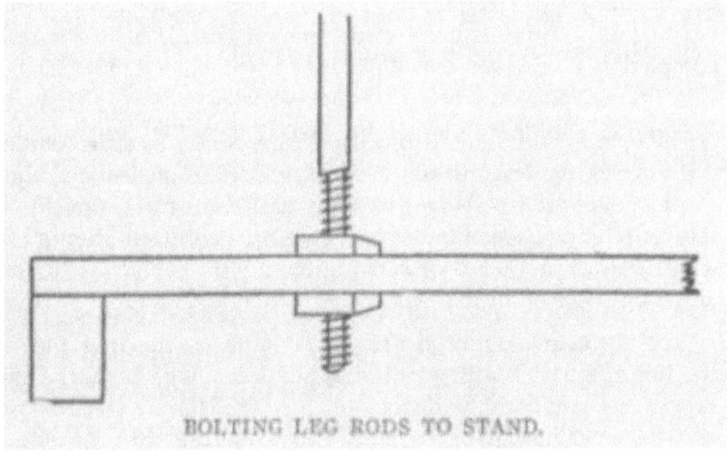

BOLTING LEG RODS TO STAND.

If two good sized rectangular holes are cut in the back-board just in front of the hind legs and behind the shoulders they will be useful later in shaping the body by sewing back and forth with a mattress needle and cord. Drive a row of lath nails into the top and bottom edges of the back-board about half their length at 2 inch intervals. They will enable you to build up first one side, then the other of the body by winding down excelsior with wrapping cord. These nails are driven fully in when the winding is finished.

The limbs also are built up by using the same material. Do not build up any part to a perfect fit yet, however, as we must leave room for a shell of paper ⅛ to inch thick. Depressions in body and limbs are reproduced by sewing from side to side or drawing down with nails.

The neck and skull are built up in much the same way and the skin fitting loosely, the manikin is surfaced up with a coat of soft modelling clay well rubbed in with a small trowel. The paper coating is to be applied while the clay is still damp so a large specimen must be partly covered with a damp cloth to prevent it drying out prematurely.

The paper for this purpose should be of some soft easily pulped variety; common building paper is good and may be torn in pieces of various size, soaked in water 15 minutes, then squeezed out and coated on both sides with paste. This is applied to the surface of the figure, the edges overlapping slightly, until completely covered. Use paper of a different color for each alternate coat to insure its completion. Five or six coats will be sufficient when it may be left to dry, after which treat it to a good coat of shellac.

The skin is withdrawn from the pickle, rinsed in soda solution, put through the benzine and meal drying and coated on the entire inner surface with preservative. Glue coated ear forms are slipped into place and fastened by long stitches back and forth through the ears. The feet and bases of ears are filled with papier mache pulp and the surface of the manikin coated with liquid glue.

Now the skin is put on the form to stay, fastening down the central line of the back with wire brads and drawn together at the junction of legs and body with stout stitches. The legs are sewn up first and the opening cut of the body last. A surplus of skin may be worked out and distributed with the point of an awl, while it may be pulled and stretched to cover a shortage in another point without changing the animal's form in the least.

The ears are pinned in place and their bases distended by tow pressed in with stuffers. Pointed wires thrust through the openings of the ears into the skull will hold them in place until dry.

The nose, lips and around the eyes are correctly placed, filling slightly between the skin and paper if necessary, use plenty of common pins to hold the skin in place. They are either drawn or cut off flush in short-haired skins when dry, but in one like the bear they may be driven to the head and left so.

Any places not inclined to stay put may be clamped down with strips of cardboard pinned on. The glass eyes should be placed now before setting away to dry, which will require some time.

When dry any bare patches of skin will have a dead appearance and require painting with oil colors thinned with turpentine to reduce the gloss. The end of the nose and lips are touched with varnish to produce the natural moist appearance.

If mounted with open mouth this is modelled in paper and wax coated as already described. The fur which should have been nicely combed after mounting will need another brushing and the animal is ready for removal to a permanent mount or pedestal. Some little judgment can be displayed in this selection as a poor, rough mounting will detract from the appearance of the best work while a specimen far below the average will pass muster with tasteful and suitable surroundings. The same principles will apply with some exceptions in mounting about all large animals.

Some of the most ponderous have a hollow wooden frame made to reduce the bulk of filling required; this is covered with wooden strips or lath and this in turn with a layer of fibrous material.

Supporting rods more than inch in diameter must have both ends threaded and be connected with the back-board by iron squares. These consist of a rectangular piece of iron, bent at right angles and drilled with a number of holes in both flanges. One set of these is for screwing to the back-board while the others are of a size to receive the upper end of the leg rod. By changing these from one hole to another it is possible to vary the distance somewhat between the front and hind legs without moving the iron squares on the backboard.

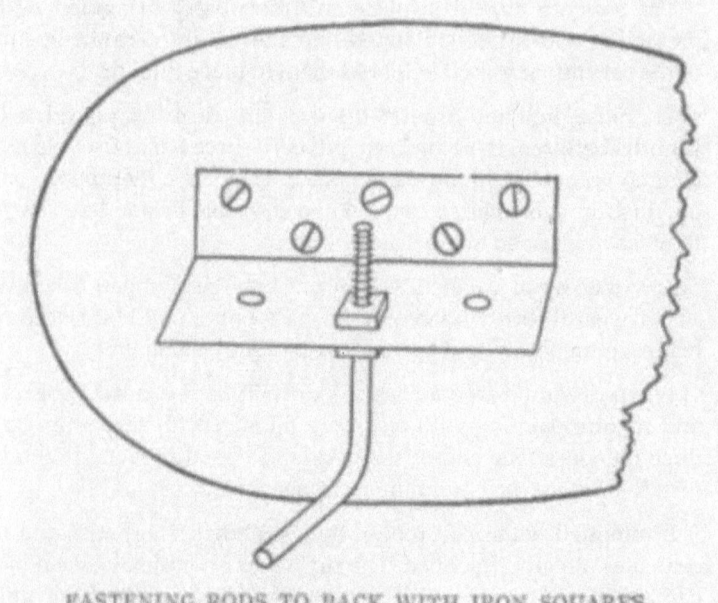

FASTENING RODS TO BACK WITH IRON SQUARES.

Sometimes the hair will be found missing in one or more places on a finished animal and in such case "Old Dr. Le Page's Liquid Hair Restorer" is the only remedy. The place to be covered is coated with glue and a small tuft of hair from the same or another skin grasped with a forceps, the base touched with glue and carefully placed. The hair is arranged with a setting needle before the glue hardens, and though a tedious operation it may be performed so well as to defy detection.

Another way where the hair or fur is of some length is to procure a patch of the right size with hair matching that surrounding, shaving the hide thin, coating the back with glue and pinning on the bare spot.

CHAPTER XVI.

MOUNTING HEADS OF SMALL ANIMALS, BIRDS AND FISH.

One of the most interesting collections which the average nature lover can make is of the heads of small game. The expense is smaller than where the entire subject is preserved, they occupy but little room, and are easily kept in good order.

WILD CAT HEAD MOUNTED ON SHIELD.

Heads of small fur bearers are all mounted in about the same way. In skinning split down the back of neck from between the ears to base of neck, cut around neck in front of shoulders and turn the scalp wrong side out over the head, put it through the usual pickling, paring, cleaning and poisoning. If ears are pocketed and lips split before pickling it may prevent the loss of hair and epidermis, in warm weather especially. Clean the skull if the head is to be mounted with open mouth. If the skull is not to be had, the teeth are broken, or you are in a great hurry, use an artificial form with the interior of the mouth already modelled.

Enlarge the opening at the back of skull and insert a piece of board not wider than the depth of neck from top to bottom. Drill a hole in top of skull and drive a screw into the board into the board inside skull cavity, prop the lower jaw open the desired distance and fill around its articulations and the base of skull around neck board with freshly mixed plaster of paris.

When this hardens the skull with open jaws is firmly fixed on end of neck board. Fasten neck board in vise and mark where to saw off, allowing for a piece of inch board shaped like a cross section of the neck. If an artificial form is used, screw it to the neck board and treat the same otherwise.

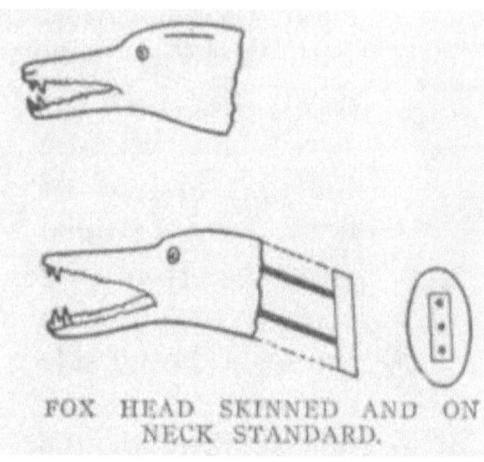

FOX HEAD SKINNED AND ON NECK STANDARD.

Make the neck short rather than long with the nose lower than the eyes in most cases. Build up neck and head by winding on tow. Mount and finish the head as directed in rug work.

The skin may be fastened at the end of neck by pins or brads driven all around the neck. Trim off any surplus with a knife, cutting from inside the skin to prevent cutting off ends of fur.

If a skin has been ripped up the front to the chin careful sewing will make it pre-

LEOPARD HEAD, AR1   HAWK HEAD.

sentable, though such seams are hard to conceal.

The heads of birds of prey and the larger game birds show up well mounted and need no special treatment from that generally given. The neck only is made up on a wire, one end of which is sharpened to thrust through the skull while the other is stapled or clinched to a bit of board round or oval shape. The skin of the base of neck is fastened to this by sewing back and forth across the back. Heads of fish like bass and pike are prepared by cutting off just back of the gills and cleaning from the back all brains and flesh.

After poisoning, fill them with tow or cotton, bracing the mouth open if wanted so and keep in the desired position until dry. Then the fibrous filling is removed and they are filled permanently with plaster or paper pulp and a piece of board fastened in the back of head to furnish a hold for screws from the back of the shield or panel.

The inside of the mouth will need remodelling with wax and the whole given a coat of white varnish. Any bright colors which may have faded should be retouched with oil colors before varnishing.

Suitable mounts for small heads are in the regular shield and round and oval shapes, and rustic panels of natural wood. A number of small heads may be mounted on one long panel.

CHAPTER XVII.

MOUNTING HEADS OF LARGE GAME.

Mounting heads, of horned game especially, is a branch of taxidermy which suffers no diminution in popularity. Such work is turned out at the present time in far better shape than it was years ago, but many fine heads still remain that were gathered in days of abundance of buffalo, elk and mountain sheep.

In skinning horned heads never open the skin up the front of the neck; not

SHEEP HEAD.

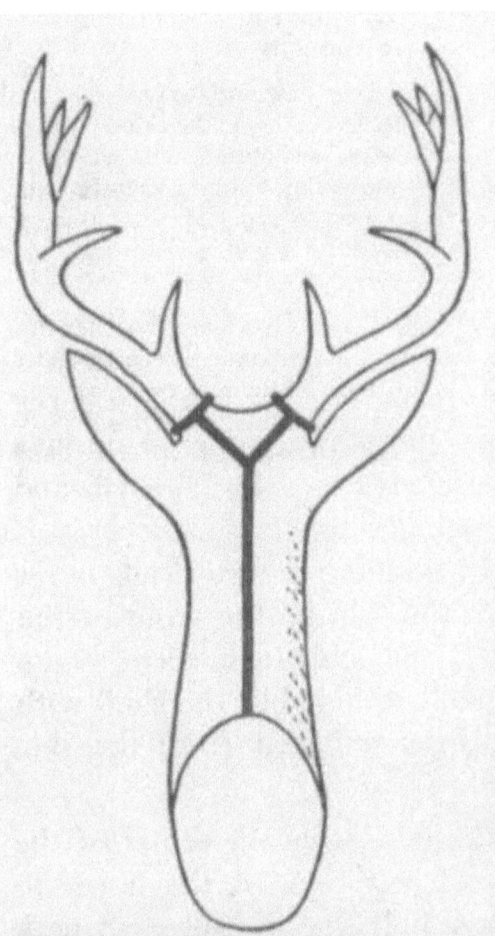

SKINNING HORNED HEADS. CUT ON HEAVY LINES.

only are such seams difficult to hide but the skull with antlers cannot be entirely removed from the skin as it should be.

To do this open the skin down the center of the back of the neck from just back of the horns to the shoulders or at least half that distance. A neck of medium length is preferable and many a fine head has been ruined by being cut off just back of the ears.

Connect the upper end of the opening cut with the base of each horn or antler by a short branch making the whole opening of a Y or T shape. Turn the neck skin inside out down over the head, which in case of a deer may rest on the antlers, until the ears are reached, cut these off not too close to the head and the horns are next met with.

Work the skin from around the base of these with a dull knife or a small screwdriver blade. With the same tool pry the thick skin away from the frontal bone. When the eyes are reached have a care not to cut their lids, working closer to the bone than the skin. Use the screwdriver again to scoop up the skin from the so-called tear pits in front of the eyes.

Let all the dark colored skin on the inside of the lips remain attached to them. The skin of the neck is the thickest on the whole animal and must be reduced by shaving. The skin of the whole head and neck should not only be freed from all flesh and muscle but shaved to about one-half its original thickness. For this purpose work on it with a sharp knife or draw shave on a half rounded beam.

Split the lips and remove their fleshy interior, split the nose cartilage and separate it from the outer skin. With some blunt tool pry the skin of the back of the ear from the cartilage and turn the ears wrong side out to their tips. Give the scalp at least 24 hours' pickling or it will be liable to excessive shrinkage on drying.

HORNED HEADS—ANTELOPE, DEER.

Many a fine head mounted green, without thinning or pickling, has shrunk and continued to shrink for months, until all stitches gave way and it cracked and shriveled to an inglorious end. If a paper head form is to be used, the top of the skull at the base of the antlers is sawn off and the balance of the skull discarded, the more common method will require the cleaning of the skull with antlers remaining on it. A little boiling will expedite this and by chopping an opening (1 inches wide in case of a deer) into the lower part of the brain cavity the brain is removed. This opening will also receive the end of a wooden neck standard of plank three inches wide.

A nail through the top of skull will hold it temporarily till the lower jaw bones are placed and the whole held solid by packing the base of skull and jaws in a mass of soft plaster which will harden in a few minutes. This neck standard should be at right angles to the greatest length of the head.

Measuring the neck skin where cut off gives the circumference of an egg-shaped board, representing a cross section of the neck at that point in a vertical line. The neck standard is sawed off at

DEER SKULL ON STANDARD.

the proper place and angle and made fast to the board by nails and screws. With a very short neck it will be necessary to depress the nose considerably that the antlers may not come in contact with the wall. This should all be calculated before fixing the skull permanently on the neck standard. The standard can be held in the vise and a little measuring will indicate the point of attachment and angle needed to clear the wall.

Now wind excelsior on the neck standard and skull until the skinned head and neck are roughly reproduced. Try the skin on occasionally to guide in this.

Do not put any excelsior on the upper part of the skull and face as no amount of flesh was removed there. Give the cheeks a natural fullness and remember the neck was not round like a stove pipe. By sew-

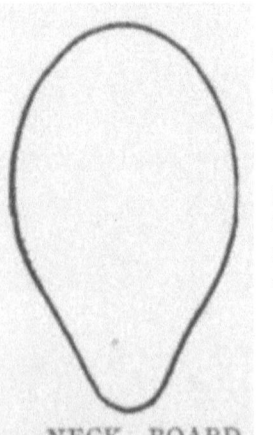

NECK BOARD.

ing from side to side the shape of the gullet and wind pipe can be molded. When the skin is still not quite filled give the head and neck a coating of potter's (or modelling) clay and then several coats of well pasted paper as directed for covering manikins for large animals.

When this has completely dried out remove skin from pickle, clean and poison it and after placing the ear forms it will be ready to cover the head and neck. I prefer good cardboard forms for the ears though some use only metal. Lead is too clumsy and heavy, copper and wire cloth corrode, pure sheet tin works nicely but is expensive.

The form should receive a couple coats of liquid glue before the skin is put on to stay. Fasten the skin in front of the eyes with a brad and draw into place about the base of the antlers. Use a heavy needle and waxed linen cord for this sewing. Heavy gilling twine doubled will do.

Sew up from first one and then the other antler to the central cut down the back of the neck, tie the threads together and continue sewing down the neck. Get the skin on face and around mouth placed, then draw the neck skin tight and nail to the edge of the board with finishing brads an inch or more long. Any surplus remaining can be trimmed away. A square of rough board, screwed on the end of the neck, will enable it to stand on the bench with nose up while the final touches are put on the anatomy of the head and face.

The split lips filled with a little clay are placed and will usually stay without pinning if the lower lips are tucked under where they belong. Fill the end of the nose and around the nostrils nicely, no live deer ever had a shriveled up nose. Fill under the eyebrows as the skin there was quite thick before paring. Set the eyes, after

PAPER HEAD AND EAR FORMS.

filling sockets with clay. A little work with a sharp awl will put the lids in place with lashes disposed aright. If the ears have not been permanently fixed, do so now filling out their bases with stuffers and thrusting stout wires into the head to keep them set until dry. A few stitches are taken to hold the skin of both sides of the ears together, when dry they are removed. Brush the hair down and if it persists in rough spots, paste them and then smooth down. When dry the comb and brush will remove the scales of paste readily along with any dirt from the hair. When dry clean the antlers and oil them lightly, brush out the hair and clean all clay from eyes and nose. Connect the eyes and lids with black wax, model the inside of nostrils with cream or pink wax and varnish the end of nose and any bare lip that may show. Pins and brads that will show are drawn out and others cut off level with the skin. The head which has hung drying on the rough board may be removed to a finished shield as complete.

FINISHED HEAD.
(Author's work.)

The paper form method has numerous advantages but is not always convenient to procure. It will save the beginner much tedious work and greatly expedite matters for the professional. These forms as supplied by dealers are of the entire head and neck. By cutting off the neck at the proper point, nailing in the neck board and screwing the plate of bone at the base of the antlers to a block in the top of head it is ready to receive the skin. It will require but a short time for the skin to dry on this foundation so the finished head is often ready to return at the end of a week.

For a number of years I have used a modification of this process. In this the form is cast in halves which are joined on a board cut to the outline of the head and neck. This will afford a secure attachment for the antlers and in addition the skin of the neck may be nailed securely each side of the opening cut, making any ripping or opening by shrinkage at that point forever impossible.

These paper forms may be bought or made in various sizes, so by the addition of a small amount of some modelling material any skin is fitted. With a supply of them on hand work can be turned out rapidly during the busy season.

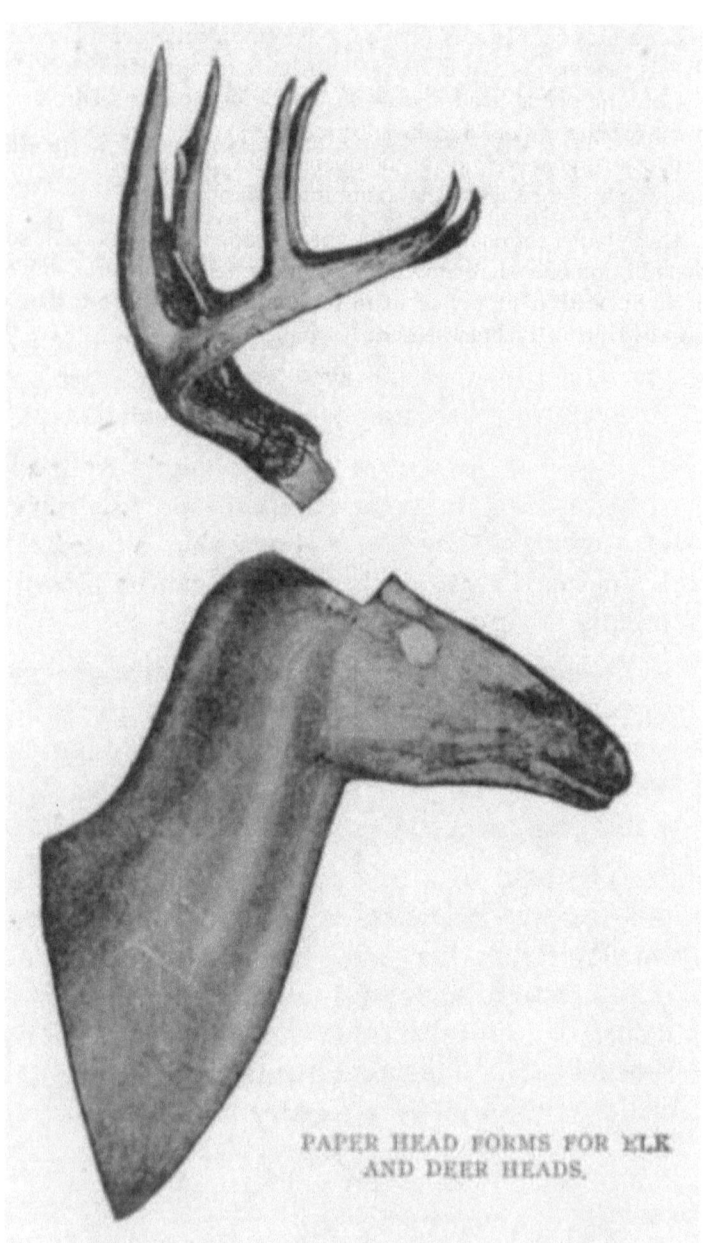

PAPER HEAD FORMS FOR ELK AND DEER HEADS.

These directions for mounting will apply equally in case of moose, elk or other large game heads, always providing supports adequate for the larger animals. A mounted head with insufficient and wabbly mechanical construction is not a joy forever.

The head of a common sheep or ram is a good one for the beginner, as its coat of wool covers small defects. It is a convenient size to handle and if not entirely successful no great expense has been incurred. On the other hand a fairly mounted ram's head is quite ornamental and suitable, especially on the wall of a country house.

CHAPTER XVIII.

MOUNTING HORNS AND ANTLERS.

A neatly mounted set of antlers or horns are an ornament anywhere, in the home, office or public room, and in case any one of the out-o'-door fraternity wishes to try setting up a pair, I will give a few simple directions and hints which may be helpful. Some bits of lumber, screws, plaster of paris, plush or leather, tacks, etc., are about all the materials needed; also a one-fourth inch drill bit to make the necessary holes in the frontal bone.

By sawing off the top of the skull down to the eyes we separate the antlers and the frontal bone on which they grow, from the rest of the skull.

Care should be taken to leave the same amount of bone on each side, so the antlers will be the same distance from the wall.

For antlers of small or medium size eastern deer, cut a heart-shaped block about 4 inches from a piece of soft ⅞-inch board. The edges of this should be slightly beveled toward one side. This may be cut out in its finished shape with a keyhole saw, or roughed out with a hand saw, and trimmed up with a draw knife or wood rasp.

After drilling two or three holes in the plate of bone attached to the antlers, arrange them evenly on this block and screw fast, using screws which will not protrude from the back of the block. If the bone is uneven or the antlers do not hang right, small pieces of wood may be inserted at one side or the other until the desired effect is had. Now put a half pint of water in some old dish and mix in plaster of paris until it is like very thin putty. With an old knife you can spread this over the bone and round it up nearly to the burr of the antlers.

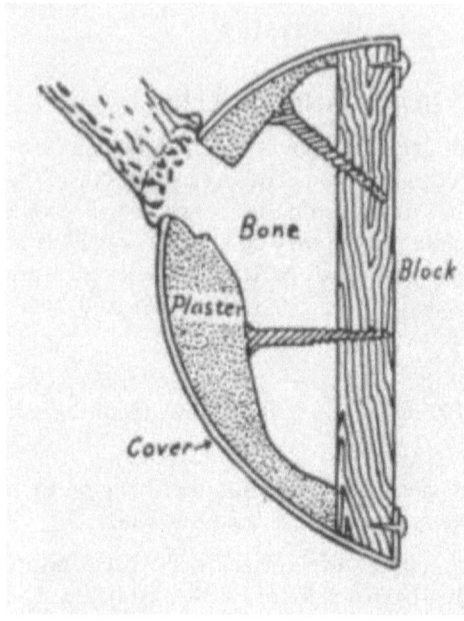

If the first mixing is not enough, mix a little more, for if too much plaster is put on anywhere it can be easily scraped off before it gets dry. This needs to be put on quickly as the plaster soon "sets" or hardens and in fifteen or twenty minutes it can be scraped and trimmed to a smooth, rounding surface.

For covering this wood and plaster base, plush, soft leather or pantasote is used. Plush or velvet is the easiest to apply for a beginner. A piece about nine inches square will do for our set of small antlers. Lay this on the plaster and turning it over the edge of the block, tack it on the back with carpet tacks, beginning in the center, at top and bottom. Slit in each side to the antler and cut a hole large enough to be a snug fit for the antler below the burr. Draw on and tack, getting the wrinkles out as you proceed, the lower, or front part, first. Lap the upper or back over it neatly at each side, turning the edges under and fastening them with a few stitches.

It is a good plan to drive the tacks only part way at first, then they can be easily drawn and re-arranged. Now cut two strips of the material to go around below the burr of the antlers. Turn the edges of these under, draw them tightly around and fasten the ends together back of the antlers with a few stitches.

They are now ready for fastening on a shield or panel. Cattle horns should have the piece of bone connecting them screwed to a long oval block, then treated similarly. Horns of sheep, cattle and goats frequently come loose from the bony core. A little plaster mixed very thin and poured inside the horn just before replacing them will fasten them on again.

Do not try to polish, paint, gild or otherwise improve the natural appearance of deer antlers. Wash and clean them well and rub in a little linseed oil. Polishing brings out the beauty of horns of cattle and bison, if the operator is lavish of elbow grease.

The process is this: Fasten the horns firmly somewhere and attack first with rasp, then file, scrape with glass, fine sandpaper, finer sandpaper, powdered pumice stone, putty powder. Finish with oiled rag. Old bison horn, weathered on the prairies till they resemble old roots, can be made to look like polished ebony by the above formula. Don't forget to add the elbow grease, though.

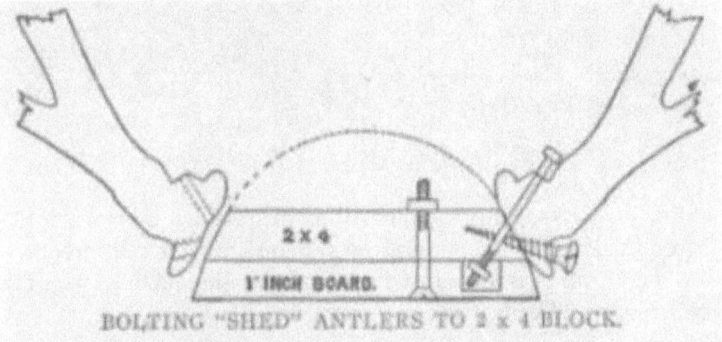

BOLTING "SHED" ANTLERS TO 2 x 4 BLOCK.

Shed antlers are a different and rather difficult proposition. It is a tedious job to drill them and insert heavy irons in their bases so firmly as to prevent turning. Often they are cut off at a bevel, drilled and screwed directly to the shield with brass round headed screws.

By drilling into the base of a shed antler, above the burr, in a diagonal direction it may be bolted to a short piece of 2" scantling. Fasten both antlers on this in a natural position in relation to each other, then drill a second hole in each and bolt them fast, using machine bolts and countersinking the heads in the antlers by chiseling.

SHED ELK ANTLERS TO BE MOUNTED. From National Zoological Park Washington, D. C.

The piece of scantling will need to be carved a little in order to get a good bearing for the butt of the antlers. This artificial forehead, as we might call it, is to be fastened to a heart-shaped block by nailing or screwing from the back and covered as directed before.

If the countersunk bolt heads are carefully modelled over with putty or "mache" and colored like the antlers no one will know they are not attached to a 'bony' fide forehead.

Elk antlers will need 5/16 inch bolts, while inch is sufficient for most deer antlers; indeed screws of that diameter will hold a small pair quite securely.

Sometimes the upper part of the skull is scraped, bleached and fastened entire to the shield with brass screws or bolts.

The base block for large deer antlers should be thicker and larger in proportion. Elk and moose antlers requiring to be fastened with heavy coach or lag-screws to a block cut from two-inch plank.

Africa has a profusion of horned game mostly of the antelope family and of late years many of these horns find their way to the walls in this country.

They are mounted as directed for the deer with the exception that many of them are improved by polishing the tips or even nearly the entire length of the horns. As most of them are corrugated or twisted in great variety this calls for considerable preliminary work with half round and round rasps and files before sandpaper, glass and polishing powders give a finish. If the tips and the higher surfaces of the balance are completely polished, the rest smoothed down somewhat and the entire horns rubbed with a little oil the effect will be good.

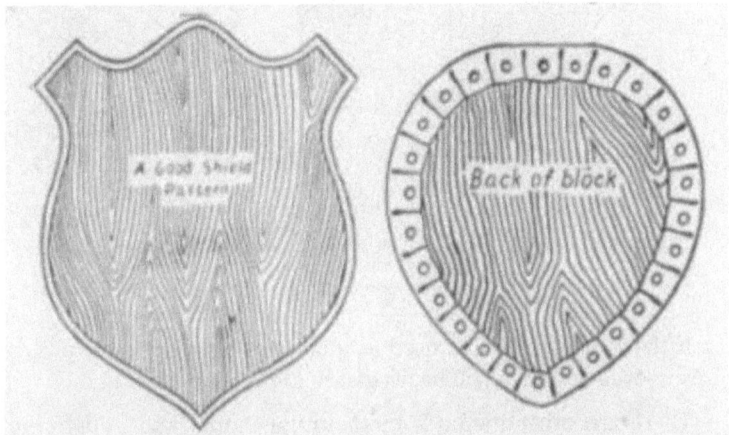

Shields are made in various patterns, woods and sizes, the average pair of deer antlers requiring one ⅞ inch thick and about 8x10 or 10x12 inches. Oak in a dull oil finish always looks well, though walnut, cherry, ash and birch are much used. If near a woodworking shop provided with a jigsaw and moulder they will turn them out in any pattern you may wish. The Ogee moulded edge is to be preferred.

If you have to make it yourself, a simple diamond, square or oval panel with rounded or beveled edge will be sufficiently difficult.

Arrange the antlers in place on the shield and mark lightly around the base, remove them and drill three holes for screws.

Countersink for the heads on the back of the shield and so fasten the antlers in place. For light horns a brass screw-eye at the top of shield is used to hang them, but heavy moose and elk antlers require an iron plate in back of shield, let in flush across the top of a perpendicular groove to catch a hook or head of a heavy nail in the wall.

DEER ANTLERS; ELK FEET; BISON HORNS.

If the antlers are to be used as a rack for hats, guns or rods, two screw-eyes or plates will be necessary to prevent turning.

There are other methods of mounting horns and antlers, but I have found the above to be the most substantial and neat, and not very difficult.

CHAPTER XIX.

MOUNTING FEET AND HOOFS.

Many sportsmen now preserve the feet of their large game to have them made up in various articles of use and ornament which they can distribute among their friends or use in their own homes. Some of these articles are gun and rod racks, furniture legs and feet, ink wells, match, cigar and ash holders, thermometers, paper weights, umbrella and cane handles.

It goes without saying that for such things as racks, furniture legs, handles and thermometer mounts the leg skin attached to the hoof should be left six or more inches in length while for ink wells, etc., it may be shorter.

WOODEN CROOK FOR DEER FOOT.

In fairly cool weather the feet and lower legs of deer will keep for some days without skinning as they contain but a small amount of flesh. Still it is safest and but the work of a few minutes to split them up the backs, skin down to the toe joints and cut them off

SKINNED DEER FO

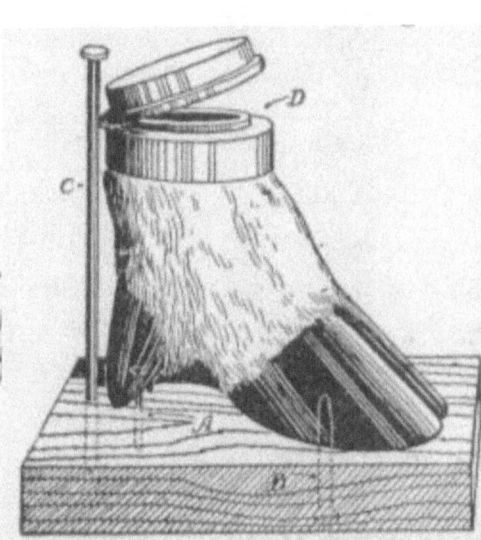

DEER FOOT INK WELL.

there. Dry them with or without salting and they are easily packed up to carry home or send to the taxidermist. If one foot and shank is received in the flesh it will aid in mounting them up as racks, furniture legs, etc., as for such purposes the skin is mounted over a piece of wood of the size and shape of the skinned leg. For preparing feet for racks and handles it is well to supply yourself with a number of natural crooks of about the size of a deer's leg and nearly of a right angle. Sassafras, gum or some soft wood work up easiest. When skinned place in pickle or give foot coating of arsenic and alum—pickle is best. Be sure and leave enough skin attached to hoof; a little experience will teach you this. Now remove foot from bath, rinse well and sew up same as far as the claws; next bore a hole through the claws from inside of claws, where it will not show. Get two wire nails and nail these claws to a board, as shown in A. Now arrange the hoofs as shown in illustration and put a screw into each from underneath, to hold them down (B), or you can nail a cleat across them by nailing to the block on each side of hoof; the idea is to get these parts firmly placed in position. Now finish sewing up the skin and stuff it full of chopped excelsior, shaping the foot as you proceed. Now drive a long nail against back side of foot to keep it from sagging (C). Allow the foot to thoroughly dry.

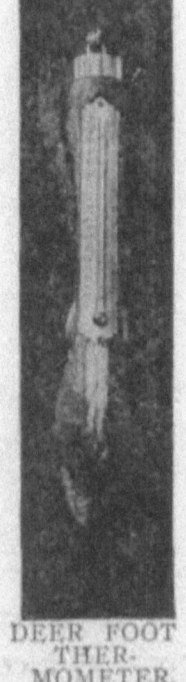

DEER FOOT THERMOMETER.

When dry remove excelsior and cut off surplus skin around top and the foot is ready for an ink well, match safe, ash tray or paper weight, as they all go on the same way. Mix up some plaster of paris in water and run the foot full and place the ink well or other fitting in place and allow the plaster to "set" and the foot is finished. If you wish a pen rest you can now place it in position. In setting up thermometers remove bone to hoof and whittle out a stick shape of bones removed. Coat inside of skin with arsenic and alum and place stick in position and sew up skin. Put on metal cap at top and tack on thermometer. For hooks on racks, work up a stick with crook into the approximate size and shape of the deer's leg with the foot bent at right angles. It had best be a little small so it can have a coating of clay or other modelling material to make the skin fit it perfectly. Sew up as for thermometer. When dry fasten to the rack by inserting in a square or oval hole and wedging at the back.

For furniture legs the feet are turned out in natural position on wooden legs, and fastened by bolting or screwing to small tables, stools, or screens. As handles for canes and umbrellas, treat the same as for hooks and leave the wood long to form a dowel which is glued or inserted in cane or umbrella, a metal band covering the end of the skin. I have referred in this chapter to deer feet, but those of elk, caribou and moose are also used and suitable fittings in nickel and silver plate are supplied in various sizes by dealers.

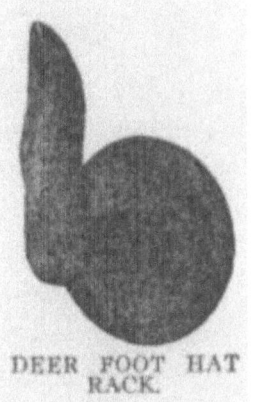

DEER FOOT HAT RACK.

If you wish the hoofs mottled (they look best that way) file same until you get to the "quick," which is light in color and gives the foot a very attractive appearance. Smooth down with sandpaper or edge of glass. Oil a rag and dip it in powdered pumice stone and rub hoof vigorously a few moments, and you will have a beautiful polish.

The smaller articles are complete as they are or may be mounted, ink wells, etc., on round, and thermometers on long panels of variously finished woods. Many nice articles may thus be made from what is usually considered worthless offal.

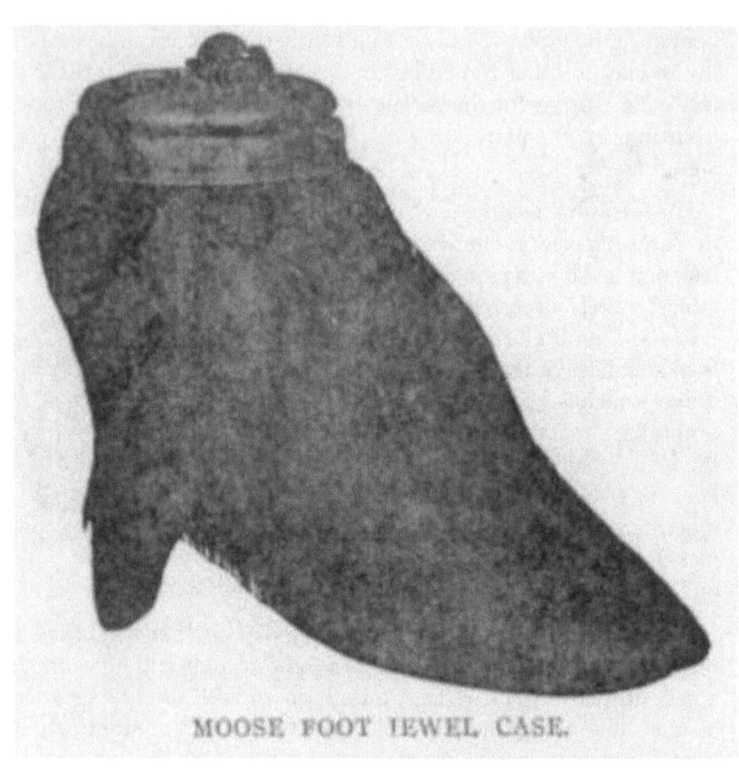

MOOSE FOOT JEWEL CASE.

CHAPTER XX.

MOUNTING FISH.

These with reptiles are most difficult to preserve with any success by the ordinary methods. There are some individuals which it is impossible for the taxidermist to prepare the skins of, so as to retain a natural appearance for any length of time. They can only be represented by casts painted to the best of the artist's ability.

Most of the varieties of medium and large game fishes can be mounted by the average taxidermist and it is with these we are mostly concerned. There are almost as many methods of mounting fish as there are operators, each having some pet kink of real or fancied superiority.

As often as otherwise fish are mounted in the medallion style, with one side only showing. This is especially adapted to display on walls and panels. For filling material everything from sawdust to plaster has been employed but as good results as any are secured by a hard core of the approximate size of the skinned fish, coated with some plastic substances which is moulded into shape through the skin.

In skinning some fish the scales must be protected by pasting thin paper over them but ordinarily it is sufficient to keep the skin wet and not allow it to dry out until it is complete. A piece of oil cloth is good to work on in skinning fish or birds either. Some taxidermists have a large pane of glass set flush in a table top for this purpose.

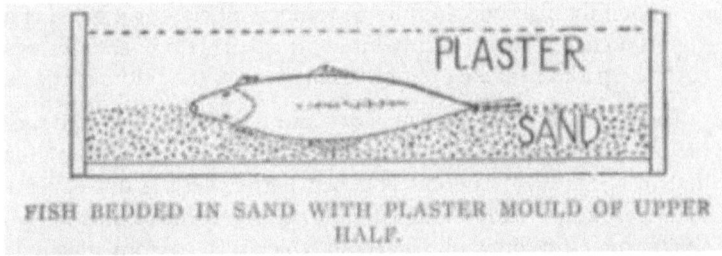

FISH BEDDED IN SAND WITH PLASTER MOULD OF UPPER HALF.

With a freshly caught fish at hand which is to be made into a medallion the process is about as follows: Before skinning lay it flat on one side on a piece of soft wood board of the proper thickness and mark out its outline. Though only one side of the mounted fish shows, a little more than precisely one half is preserved in order to include the tail, the vertical and dorsal fins and also to give an appearance of rotundity.

Leaving this margin all around, the skin is cut away from the side which, on account of damage or other reason, is selected for the back. The head must be cut through with chisel or a fine saw. Severing the bones at the base of the fins with the scissors the whole body is removed by cutting the spinal column at its junction with the skull.

In skinning fish you will note a white layer, like tin foil, which gives the fish its silvery color. Do not disturb this if possible. Remove all surplus flesh, cut away the gills and interior of head and if at all greasy (what fish is not) treat to a bath in gasoline. Use absorbent, sawdust or meal to remove oily gasoline, drop in alcohol or formaldehyde solution while the body is prepared. To do this cut out the board by the outline on it with a short bevel on the back and the other side the full shape of the fish. The whole form is slightly diminished, however, to allow for a coating of clay. This is applied after the skin has been removed from the solution and poisoned.

When a good fit is secured the skin is fastened in place by sewing from top to bottom, across the back board, or if large, nailing the edges to the board. The fins which have been kept wet are to be spread; each clamped with two pieces of pasteboard. On very large fish spring clothes pins may be used to clamp the fins, for small ones pins forced through both thicknesses, *outside* the fins.

Sponge the fish off carefully to remove all clay or other dirt and give it a coat of rather thin white varnish. This prevents the scales curling up and to some extent fixes or restores the colors of the fish.

The eye is set after the fish is dry and if it does not fill the socket, model around it with wax or paper pulp. Fish eyes vary so greatly that to strictly copy nature you had better use the uncolored fish eyes, painting the back with suitable oil colors with a coat or two of shellac over it to prevent the clay in which it is set from affecting the paint. The final painting of a mounted fish which is necessary to complete the best work is a task for an artist. If a specimen in the flesh (living if possible) is at hand this is made easier. All fish skins collected should be accompanied by color sketches if possible.

All silvery fish should be coated with size and nickel leaf over their entire scaly surface. On this ground paint with thin oil colors. If the paint is not too thick the desired silvery sheen will show through. If the whole fish is dark no leaf is needed and in some cases the upper part of the body requires a gold ground with the nickel leaf on the silvery under parts. Japanese gold paint or something similar will do for a golden ground.

The finished medallion may lie flat in a case, be fastened on the face of a panel, or hung by a loop at the mouth or center of back. Panels of natural wood are a favorite mount and framed panels covered with plush or the imitation pebbled upholstery leather.

Another method of mounting medallions is to take a plaster mold of the display half of the fish and from it make a plaster cast like the back board. This is sandpapered down to allow for the skin and gouged out at the bases of the fins and tail. The head too is not reproduced on the cast.

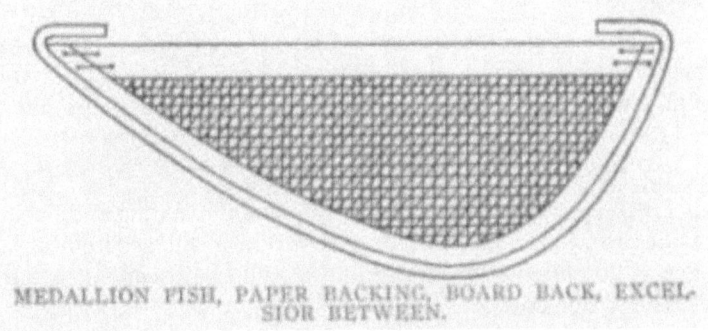

MEDALLION FISH, PAPER BACKING, BOARD BACK, EXCELSIOR BETWEEN.

When the skin is ready to apply, coat this plaster cast with some nonporous modelling material. A mixture of thin liquid glue and whiting is good for this. Some paper pulp is put inside the head and at the junction of body and fins. Shape from the outside with the fingers.

A piece of wood should have been placed in the back of the plaster cast when making same, in order to fasten to a panel by screws from the back. Paint the wood with melted paraffin before putting in the wet plaster or it may swell and subsequently shrink enough to crack the cast. By either of the preceding methods the entire fish can be mounted if desired.

The opening cut should be made from head to tail along the lower edge of the body in most scale fish and will require some neat modelling to hide as both sides are to be on exhibition.

Entire fish are usually supported by metal standards rising from a wooden base. Such standards are preferably of brass, threaded and fitted with nuts and rosettes at both ends. Two nuts at the lower end clamp the base and with the upper end inserted in the back board, the upper nut will adjust the fish at the right heighth. The rosettes cover the nuts and add a finished appearance.

Especially adapted to tarpon and other ponderous fish are medallions mounted on paper. To do this a half mould is made as described, the skin removed, cleaned, poisoned and replaced in the mould. Then it is backed up with numerous layers of soft paper, well pasted and pressed in with the hands. Let these layers of paper overlap the mould at least as much as the margin of skin left on the back.

When a sufficient thickness is attained fill the hollow form loosely with crumpled paper or excelsior and fit in place a back board of light wood. Nail through the margin of skin and paper into the edges of this. If a number of large holes were made in the back board it will expedite the drying out.

When partly dry, remove from the mould, painting and finishing as before when completely dry. By this means the contour of large fish is absolutely reproduced and the finished work is extremely light and durable.

Many of the smooth skinned fishes are impossible to mount satisfactorily, a cast is the best we can do for them.

When using a living fish as a guide in painting, it may be confined in an aquarium and by sliding a wire screen partition, be kept just before the artist's eye.

CHAPTER XXI.

MOUNTING FISH—BAUMGARTEL METHOD.

Some years ago A Practical Method of Fish Mounting was advertised by Mr. Baumgartel in Angling and Sporting publications. Entire satisfaction was given to those who studied and applied the lessons, through correspondence school methods. Both the author and publisher of HOME TAXIDERMY FOR PLEASURE AND PROFIT, are indeed glad to publish the entire course as used by Mr. Baumgartel, including diagrams, figures, etc., as same together with copyright was conveyed to A. R. Harding.

The same degree of excellence in mounting fish has not been so generally attained as in other branches of the taxidermy art, and this I believe is because an equal amount of study has not been given the subject. Hundreds can mount birds well to one who can prepare fish in as skillful a manner, although the mounting of fish dates from as early a period in the art of taxidermy. It is a question of method and the right one.

The usual methods of mounting fish have proven so unsatisfactory that they have been almost entirely abandoned, and, until the method to be described was devised, it was necessary to place specimens in alcohol or other preserving liquids or to make plaster casts of them. The objections to the former process are that it is expensive, requiring especially constructed jars to show the specimens without distortion. They usually lose all their natural colors, and in most cases shrink to such an extent as to give only an approximate idea of their original form. There are also serious objections to the latter method. Plaster casts are easily broken and certain parts, such as the interior of the mouth, cannot readily be produced. Further, it is not desirable in natural history collections to exhibit casts of the objects when the originals can be displayed. Then, too, the sportsman does not care for a cast of his "Big Fish," but wants the real thing to verify history of the one that didn't get away. To me a plaster cast of an object that can itself be preserved is about as interesting as a nicely painted decoy duck compares with a well-mounted skin.

As is the case with nearly every taxidermist, professional as well as amateur, I have always been an enthusiastic sportsman. The desire to preserve the specimens taken by me led me to devote myself to the study of taxidermy. Perhaps the dilatoriness of the "artist" who mounted my first specimens had a stimulating effect. Later, as a professional taxidermist, I for many years mounted fish by means of the various methods, but the results obtained were not satisfactory to me.

In 1903 Mrs. Baumgartel and I made a trip to Pine River, Michigan, for trout. It was in June. The weather turned cold, and we took few fish. On the twelfth I made what was to be my last cast, had taken down my rod and was walking along the bank of the river on my way to camp, when at the edge of a pool I noticed a fish jumping. I could not resist the temptation to try one more cast, and making preparations I dropped a Parmachene Belle a few inches from the spot where the fish had just broken the water. There was a rise, a strike, and I was fast to a fish destined to be mine. After an exciting struggle, I landed a thirteen-inch grayling weighing one pound and two ounces. Of course, this fish I had to preserve and wanted it to look as it did when taken from the net. We boarded the train for home that evening, and supplying my wife with reading matter, I was soon lost in meditation. The madam told me many times during the journey home that I was not at all companionable. Be that as it may, the result of the earnest study given the matter, and the previous experience gained in experimenting with other processes, was that a practical method of mounting fish was devised, a method by means of which the angler can successfully mount his own specimens. "Necessity is the mother of invention," you know.

A picture of this grayling is given below. It still occupies a position in our dining room, together with others to remind us of pleasant days on lake and stream. It is mounted with the very fly and leader on which it was taken.

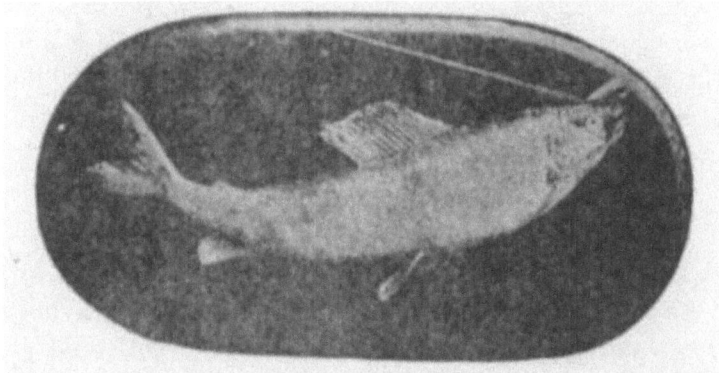

GRAYLING—RESULT OF MY FIRST EXPERIENCE AT FISH MOUNTING.

For the past four years I have successfully taught my method by correspondence. A few years ago activities in other directions compelled me to give up taxidermy work, and other interests now demanding more of my personal attention, the publisher of this book, always on the alert for something of practical value to interest his readers, will present to you in these pages the identical instructions I have so successfully used. I take the liberty of quoting verbatim from a letter just received by me from a friend of a prominent eastern professional man, one who, while "chained to business," still finds time to get "close to nature" for a season each year:

"Referring to the fish mounted on our Newfoundland trip: I may say that it was our first experience and we were agreeably surprised at results. In all there were ten specimens of salmon, sea-trout and brook-trout mounted and we found no great difficulties. All the work was accomplished in an old barn after we got home evenings and early mornings before going out. The only obstacle we encountered was getting the fish back to the states in good shape. Five of the specimens now occupy prominent place in the Doctor's den and I am always pleased to point out to friends the results of our labors."

You can do as well.

INSTRUMENTS.
Knife.
Needle and thread.

Saw, fine toothed.
Scalpel.
Scissors, straight and curved.
Shears.
Skin scraper, not the toothed-edge.
Tweezers.

One can get along with simply a jack-knife, pair of shears, and needle and thread; but to do first-class work easily, good tools are required.

MATERIALS.
Alcohol.
Aluminum leaf.
Arsenic, powdered.
Clay, potters' or modelling.
Eyes, glass, clear except pupils.
Papier mache, prepared.
Pine board.
Pins.
Plaster, calcined.
Tube paints.
Varnish, clear white.
Brushes.

PRELIMINARY INSTRUCTIONS—Try first a perch or other fish having scales firmly attached. See that the fins and tail are uninjured, and that no scales are missing from the side to be displayed. As every perfectly formed fish has both sides alike, and as ordinarily but one side can be seen at a time, only a little more than one half of the fish is to be mounted.

Note carefully the colors (including those of the eye and the interior of the mouth) of the freshly caught fish, making a sketch of the specimen, showing the extent of the different colors and markings, the spots, if any, and the eye. The pupil is not usually round. The eye of a lake trout appears like Fig. 1. A carefully kept note book is a valuable aid.

While the tail is of course a fin (the caudal), in this work it will be called the "tail," to distinguish it from the other fins. See Fig. 2 for key to the names of the fins.

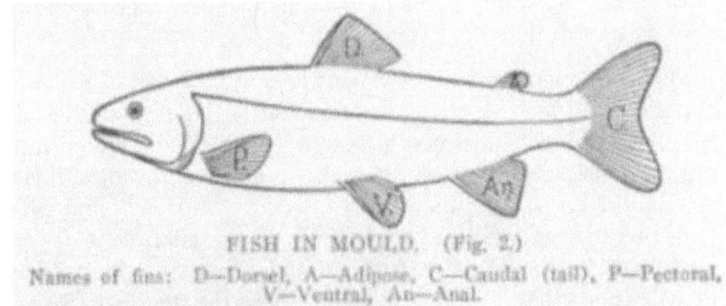

FISH IN MOULD. (Fig. 2.)
Names of fins: D—Dorsel, A—Adipose, C—Caudal (tail), P—Pectoral, V—Ventral, An—Anal.

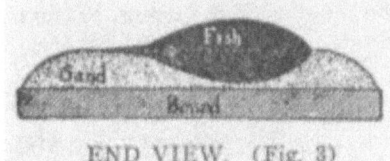

END VIEW. (Fig. 3)

THE MOLD.—Wash the fish in water to remove dirt and mucus. On a board somewhat longer and wider than the fish, place a sufficient amount of moist sand in which to imbed the specimen to one-half its depth when lying on its side in the desired position. Level the sand and hollow it out for the larger part of the body. Large fish should be displayed straight, with the mouth closed or only slightly open. Smaller and more graceful ones may be shown in positions of activity; rising to fly, breaking away, etc. If the mouth is to be open, fill it with cotton or cloth in order to keep out the plaster. Place the fish in the position you wish it to have when mounted, the side to be displayed uppermost. See that the dorsal and anal fins, and the tail, lie flat on the sand, and that one-half the body appears above the sand its entire length. There must be no uneven places. Viewed from the end, it should appear like Fig. 3; from the side, like Fig. 4.

SIDE VIEW. (Fig. 4)

Heap up the sand all around, about an inch from the fish, to prevent the plaster flowing off the board.

Mix a sufficient amount of plaster to cover the fish to a depth of about half an inch, covering the fins and tail as well as the body. Mix the plaster by stirring a little at a time into cold water until it has the consistency of cream. Place the pectoral and ventral fins flat against the body. Pour the plaster over the fish slowly and evenly (covering the head, tail and edges first), allowing it time to dry until quite hard, perhaps thirty to forty minutes. Then turn the mold over (it will appear like Fig. 2) and, by inserting the fingers in the gills, carefully remove the fish. Lay the mold aside for a time. Wash all sand and plaster from the fish.

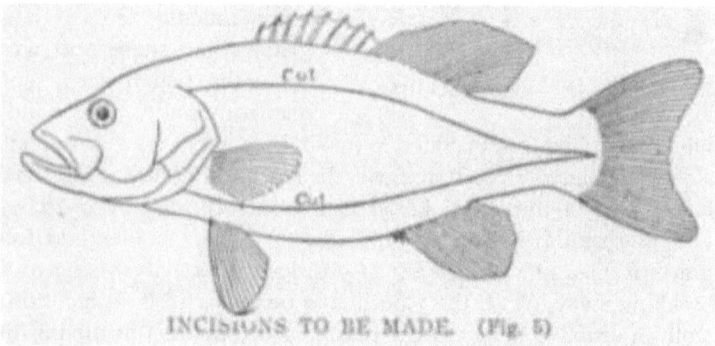

INCISIONS TO BE MADE. (Fig. 5)

SKINNING THE FISH.—If the fish has scales easily dislodged, wrap it, with the exception of the fins and tail, in several thicknesses of tissue paper, which will readily adhere to the moist skin. Lay the fish on your table, on the side which was covered by the plaster and place wet cloths on the fins and tail to prevent drying. Commence at the gills and make two cuts with the scissors or scalpel lengthwise of the fish to the tail, cutting into the abdominal cavity below, joining these cuts (see Fig. 5), removing the strip of skin, including the pectoral fin, with adhering flesh, and the contents of the cavity.

With the fingers, or tweezers, grasp the cut edge of the skin of the back and with the scalpel carefully separate it from the flesh as far as the middle line of the back from the head to the tail. Remove the exposed flesh to the backbone. With the knife, shears or fine tooth saw, split the head lengthwise a little to one side of the middle, leaving somewhat more than half. Do not sever the skin of the body where it comes to a point between the gills, and use great care when removing the flesh from this portion.

You now have a trifle more than one-half the fish to work on. With the scissors cut through the ribs and remove the backbone with some of the flesh. Be careful when cutting through the backbone at the tail not to cut through the skin below. Going back to the head, remove the remaining flesh, and with the curved scissors and scalpel cut away all the cartilage possible from the head; the more the better so long as the skin is not injured. If enough of the cartilage can be removed to expose the muscles of the cheek from the inside, cut them away, taking out the eye; otherwise it will be necessary to work from the outside of the skin through the eye opening, and this must be done very carefully or the skin of the cheek may be broken. With most fishes it is possible to remove all the cartilage from the head, and this should be done to prevent shrinkage. If the mouth is to be open, do not cut away its lining or much of the tongue behind. The tongue is to be split perpendicularly lengthwise and about one-third of it removed. When the head has been thoroughly cleaned, remove the remaining flesh from the body with the skin scraper or scalpel.

The ventral fin which is uppermost as the fish lies on its side and is not to show, should be carefully cut off outside the skin. Do not cut off the ventral fin on the side which is to be displayed. Do not scrape away the silvery lining of the skin if this can be avoided. Some of it will come off. Cut away the bases of all the fins and the tail inside with the curved scissors and scrape away all flesh, working close to the fingers so as not to stretch the skin. Tie the vent inside with thread. Unless the fish is quite small, the skin of the lower jaw must be loosened with the knife or scalpel and the muscles cut away. The adipose, or small fleshy fin on the back near the tail of such fishes as the trout, must be carefully opened from the inside of the skin with a small-bladed knife and the contents removed, to be later replaced with clay.

Place the skin in water to loosen the paper—if any has been used—carefully washing the skin and wetting the fins and tail thoroughly. You may allow the skin to remain in the water until you are ready to put it back in the mold, but not longer than a few hours.

FILLING.—Brush the sand from the mold, and if the upper edge is uneven, smooth it with a knife so that it will be perfectly straight. Should there be any rough places on the inside of the mold, carefully scrape them down with the skin scraper.

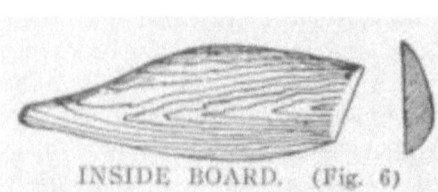
INSIDE BOARD. (Fig. 6)

Cut out roughly a piece of soft pine one-half to one inch thick the shape of the outline of the fish, but somewhat smaller, using the mold as a guide. One side will, of course, be flat, and that side should be uppermost when placed in the mold. Work up with water a sufficient quantity of clay to about the consistency of fresh putty. Place the mold on your bench or table, resting it on something soft (such as a piece of old carpet or burlap) to prevent its breaking. Drain the water from the skin and put it back into the mold, adjusting it nicely. The median line will guide you. See that the head, fins and tail occupy the same places they did before. Pour a little alcohol on the skin inside and let it run along the bases of the fins and tail, over the entire inner surface of the skin and into the head to preserve any bits of flesh that may possibly remain. Drain off the surplus alcohol. Fill the adipose fin (if the fish has one) with clay. Sprinkle powdered arsenic over the entire inside of the skin and head. Do not use more than will readily adhere.

The chances are that you removed more or less of the silvery lining of the skin. Whether you did or not, cover the entire inner surface of the skin of the cheek and body with two or three thicknesses of aluminum leaf. Do not cover the entire surface at once. Cover a small part at a time, and then put on top of the leaf enough clay to cover it, commencing at the head and continuing to the tail. Replace the muscles and cartilage of the head with clay. Be sure to keep the skin properly adjusted to the mold. See that the fins and the tail remain in their proper places and that they are kept covered with wet cloths.

Flatten out the clay (a small quantity at a time) with the fingers and cover the inside of the skin to the depth of about one-eighth of an inch, pressing it down firmly, especially at the bases of the fins and tail and into the cheek and head. Fit in the piece of pine, cutting it down as may be necessary. Its shape will be something like Fig. 6.

Mix some more plaster, and before putting in the piece of board, pour in the plaster on the clay, filling the skin perhaps two-thirds full. Quickly, before the plaster sets, put in the piece of wood, carefully pressing it down into the plaster until it is level with the upper edge of the mold, removing any surplus plaster quickly and neatly. Sew up the skin from one edge to the other as shown in Fig. 7.

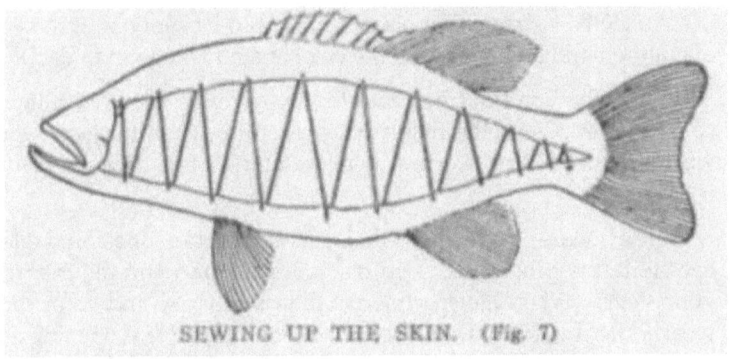

SEWING UP THE SKIN. (Fig. 7)

REMOVING SPECIMEN FROM MOLD AND DRYING.—Lay a piece of clean board of the proper size on top of the mold, turning both over. Lift up the mold a trifle, gently shaking it. The fish may or may not come out. If not, turn the mold back, insert the point of a knife in the wood and try to start the fish. In extreme cases it may be necessary to break the mold carefully. However, there should be no undercuts to hold the fish.

Carefully wash the fish as it lays on the board to remove all clay and plaster which may be on the upper surface. With the fingers smooth out any wrinkles or uneven places. Sometimes, when drying, small wrinkles or bubbles may appear in the skin of such fish as trout, but they will soon disappear. If you placed the fish in a natural position when making the mold and properly adjusted the cleaned skin to the mold, there should be no wrinkles.

Place thin pieces of wood the shape of the fins between the board and the dorsal and anal fins, which should stand out from the board a little way. Cover the tail and the fins which lie flat with thin pieces of wood, pinning them to the board until dry. The fins which do not lie flat should be spread between thin pieces of wood held in place with pins or clips.

When the surface of the fish is dry, which will be in from six to twelve hours, give it a coat of the very best white varnish (the ordinary yellow varnish will not do) and put the specimen in a well ventilated place out of the sun to dry. In three or four days when the fins are dry, remove the thin pieces of wood and apply another coat of varnish.

CAUTION.—Arsenic is poison and should be kept out of the way of children and animals. Keep the box covered when not in use.

Cuts in the hand can be protected by covering them with liquid court plaster. Clean the finger nails carefully when through work, washing the hands in warm water containing a few drops of carbolic acid.

FINISHING.—While the fish is drying secure uncolored glass eyes with the properly shaped black pupils and paint the iris from your sketch. When the specimen is thoroughly dry, in two or three weeks, dig a sufficient amount of clay out of the eye opening and put in the glass eye, setting it in papier mache. Use the prepared mache, which only requires boiling with water for preparation. When the mache is dry, give the exposed portion of it one or two coats of paint of the proper color.

Now do such painting as may be necessary—for instance, the spots and fins of the brook trout, colors of which have doubtless vanished by this time. Use tube paint, thinned with the white varnish. Usually it is sufficient to place a small quantity of the paint of the proper color directly into the varnish. Do not use much of the paint—just enough to secure the color and yet not obscure the scales. Where the markings are prominent, put some of the paint directly on the fish and spread it with the varnish. Brilliant spots, such as those of the trout, can be reproduced by the use of the paints without the varnish.

While the specimen is drying prepare a panel for it. To show the fish to the best advantage, the panel should be of polished hardwood, although stained pine will answer. Bore two holes about half way through the panel from the back, slanting upward, by which to hang it. (See Fig. 8.) Bore two holes entirely through the panel in the proper places and screw the fish to it, putting in the screws from the back of the panel and into the fish where the wood is thickest. Countersink the screws.

HOME MADE PANEL. (Fig. 8)

Finally, apply a last coat of the varnish. Do not varnish the glass eye. By keeping a piece of writing paper between the panel and your brush you can varnish the fish without getting any on the panel. It is best to put on the final coat after the specimen is mounted on the panel, because if the fish is handled before the varnish is hard finger marks will show.

SIDE VIEW. (Fig. 9)

MOUNTING HEADS.—With a sharp knife or saw cut the head off squarely just back of first (pectoral) fins, as shown in Figs. 9 and 10, which show the head of a black bass. In this case the ventral fins are also left on. Place the head on a board with the cut part down, spreading the fins as in Fig. 10. If it is to be displayed with open

159

mouth, fill the mouth with cotton or cloth to exclude the plaster. Cover the whole head with plaster.

After the plaster sets, with a saw and knife cut the mold into two parts lengthwise, being careful not to cut into the head. Use the saw first and when the plaster is cut down close to the skin use the knife carefully. Do not attempt to remove the head before cutting the mold in two.

Remove flesh and cartilage from the head,

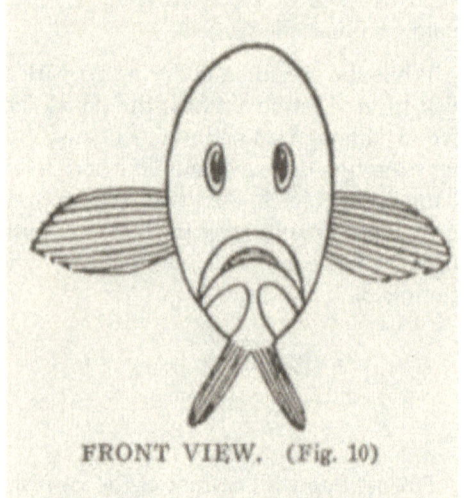

FRONT VIEW. (Fig. 10)

line with aluminum leaf, and proceed as previously instructed. Of course none of the tongue is to be removed if the mouth is to be left open, in which case do not remove the bony parts of the gills. Before placing the cleaned skin in the mold, tie the two parts of the mold together. Cut a neck board to fit and set it in plaster. Finish as previously advised.

CHAPTER XXII.

MOUNTING REPTILES, FROGS AND TOADS.

The largest reptile of the United States, the alligator, is mounted by methods applied to medium sized animals. Leg, head and tail rods are stapled to a stout back board and after building up the legs from tow the larger part of the body is filled by stuffing with coarse tow or fine excelsior. Let the skin rest back down while engaged in this, sewing up the skin as it proceeds, with stout twine and a sail needle. You may even need to use the awl to pierce the armor like skin.

For any natural position the leg irons need not be heavy as this animal usually keeps its body and tail in contact with the ground. The leg rods are clinched or bolted beneath the pedestal as in other quadrupeds and in addition some long screws are turned into the back board from below and the tail held down by wire fastened to its central support and clinched beneath the pedestal.

All but the smallest lizards are mounted in the same way as the small furbearers. There is apparently no known mode of "stuffing" a snake so as to resemble its natural state.

The skin must be placed on a carefully modelled manikin with a plastic layer between. For small snakes tow is wound on a wire and shaped with thread, and excelsior is used in the same way for the large ones.

The larger manikins are to be posed and paper coated in most cases before receiving the skin. Frogs and toads are also very difficult to mount in natural positions, but are nicely represented in painted casts.

Frogs, however, possess the distinction of not having to be sewed up, when skinned as they usually are through the mouth. In doing this the entire body is dissected away through the mouth and the legs are detached and skinned the same way.

WIRING SYSTEM FOR FROG.

After turning completely wrong side out and poisoning the legs are wired, wrapped with tow or cotton in the same manner as other small animals. One hind leg wire is cut long enough to reach through the body and head and to this the other leg wires are twisted. Some claim that to leave the vertebral column attached to the skin of the back is an invaluable aid in giving that part its proper shape.

The body filling is tow or cotton placed through the mouth in small pieces until the proper shape is acquired. Dry sand has been used to fill the bodies of frogs, being poured in the mouth through a funnel and retained by a cotton plug until the skin was dry, when it was poured out.

Painting and varnishing are required to finish mounted frogs. The frog is a favorite with the caricaturist as it can be made to take almost any human posture with laughable results.

Turtles may be mounted by wrapping and wiring legs, tail and head like other small animals, after detaching the under shell on three sides, removing the body and skinning the limbs. The tow wrapped legs should have a covering of soft clay which can be shaped with the fingers after they are returned to the skin.

Twisting the wires together is all right for the small turtles, large ones need a block of wood to clinch the wires in. The under shell is replaced and fastened with small wires and as enough skin was left attached to it to sew to, all cuts are closed that way. Heavy wires are seldom necessary in turtles. Those having bright colored shells will need to be touched up with paint and all should be varnished thinly to give a fresh appearance.

CHAPTER XXIII.

SKULLS AND SKELETONS.

While the preparation of skeletons for the cabinet is sometimes undertaken by the general taxidermist it is more often the work of a trained osteologist. Collectors in the field are often asked to preserve rough skeletons of desirable varieties and the skulls at least should be preserved with the skin of each quadruped taken for mounting.

A specimen with a damaged or imperfect skin may yield a good skeleton and in the case of something very rare both the skin and skeleton may be mounted separately. This process is one calling for a skilled operator as all claws, nails or hoofs should remain on the skin while their bony cores are part of the skeleton.

In the preservation of rough skeletons, skinning by any method is the first step, next the removal of the viscera, etc., then the most of the flesh and muscle should be dissected off the bones, after which poison with dry arsenic and put where it will dry out quickly and be out of the reach of foraging animals.

The legs of small animals should be unjointed as well as the skull and after trimming be put inside the body cavity and securely tied to prevent loss; birds are treated about the same and all large animals are pretty thoroughly taken apart in order to properly clean the bones.

Always remember that a skeleton with parts cut away or bones lost is about as good as none. Leave any cartilage attachments and any parts of a bony nature for the osteologist, to be on the safe side. Sometimes along salt water an uncleaned skeleton may be put in a wire netting cage and anchored in the water where various small marine animals will soon clear away the flesh. On land, too, a similar expedient may be practiced by putting small carcasses in a box with holes bored in it and burying it in some active ant hill. In both cases the openings need to be small, that the smaller bones may not be carried off and they should be removed before the ligaments are destroyed.

SKULLS—DOG WOLF, SHE WOLF, BAY LYNX, OTTER, MINK.

When they are not wanted for scientific purposes, skulls may be cleaned with the minimum labor by boiling. Watch them closely, however, and remove as soon as the flesh gets tender as much cooking will cause the teeth to fall out and the skull to separate at the sutures. Glue and plaster paris will put such disintegrated skulls in shape for commercial mounting but they are ruined forever for the scientist.

A friend was once cleaning a quantity of skulls (for museum purposes) and to expedite matters put them on to boil; all went well as long as the pot was watched, but an accident, the collapse of a large building, called him away and prevented his return until a dozen or so skulls had turned to a mass of loose teeth and scraps of bone. I never knew just what transpired between him and the museum curator afterward, nothing of interest to the general public.

Small specimens which it is proposed to skeletonize are best preserved entire in alcoholic solution as loss and breakage are thus prevented. The solution of formaldehyde can be used for this purpose but is not as good as it toughens the flesh, making its future removal more difficult. The complete cleaning of a skeleton is a matter of much soaking and scraping, calling for much patience and a strong stomach. Ligamentary skeletons of the smaller birds and animals are often prepared and mounted by the non-professional with fair success.

The entire specimen is cleaned of all flesh without disconnecting any of the bones except the skull and the leg of all but the smallest species.

The spinal cord is replaced with a brass or galvanized wire of suitable size and length; this should project enough to penetrate a piece of cork fitted to the cavity of the skull. If the leg bones were removed they should be fastened back in place by drilling small holes through them at the joints, inserting a piece of brass wire and clinching the ends over.

The skeleton is hung by cords or threads in a frame of wooden strips, so the feet will rest on the base, and then arranged in some natural attitude, holding the parts in place until the ligaments are fully dry by means of pins, threads and strips of cardboard.

The finished skeleton had best be supported on its pedestal by two metal rods with a U shaped fork at their upper ends which will clasp the vertebrae just in front of the hind legs and back of the head. These rods should be of brass or galvanized iron gilded and their lower ends are either threaded and provided with two nuts, or bent at right angles and stapled to the under side of the mount. Bird skeletons are treated in a similar way, but the wing bones need a supporting wire fastened to the back bone and a single standard. The smaller birds and animals up to the size of a small squirrel may be skeletonized and mounted without metal supports.

A ligament which gives way may be replaced by some fibres of raw cotton saturated with glue. While cleaning the bone for a ligamentary skeleton it should be kept damp until it is given the final attitude. Water with a few drops of carbolic acid should be used for this. A bath with chloride of lime solution will help to whiten the bones, though very greasy ones call for an application of benzine.

Fish, reptiles, etc., demand about the same treatment. The large birds and quadrupeds are usually cleaned bone by bone, and each joint articulated in the laboratory, though their preservation in the field as rough skeletons require similar methods.

The main rule in collecting skeletons is to never, never lose a bone or anything of a bony nature attached to the specimen.

CHAPTER XXIV.

SPORTSMEN'S TROPHIES.

As our game becomes scarcer I believe there will be more demand for the preservation of the sportsman's trophies than in the days of abundance now past. Then only a phenomenally rare or large or freakish example seemed to warrant the trouble and expense of putting in the taxidermist's hands. Now the souvenir of a good day's sport or a memorable outing is deemed well worth keeping.

Heads, horns, skins for floor or hangings and fish and game panels for the dining room walls have always been in high favor with sportsmen. So also are unique articles of use and decoration for the home. The naturalist sportsman whose trips are, from force of circumstances, only local can in a short time make a splendid showing by preserving such good types of game as he may procure.

In mounting birds as hanging dead game it is well to hang the specimen before skinning, in the position wished and if possible sketch it so, at least impress its appearance well on the memory. The main points of the process are the same as for ordinary mounting. There are, however, a few exceptions which I will mention.

If one side of a bird is defective in any way it may be mounted with such side next the panel, so often, if the specimen is to have the breast or under side displayed, the opening cut is made down the back or on one side. If a pair of birds of the same kind are used on one panel pose them to display the back of one and breast of the other.

It will usually be necessary to wire the wings of birds for game panels so as to adjust them properly, though they are sometimes fixed from the outside by embedding sharpened wires in the body.

Ducks of all kinds are especially suitable for panels in that their plumage being stiffer and more durable does not make casing in glass so necessary, though most of our game birds can, by proper treatment, dispense with such protection. One of the most effective duck trophies which I ever saw was a string of three or four small duck rising in flight apparently from one corner of a room, to the ceiling in the center of the side wall.

For this effect they are mounted with wings spread and raised, head and legs outstretched. They are hung on nails in the wall in a regularly ascending line, the point of suspension being a wire loop under the wing on the side next the wall. Single birds look well in the same position. Rabbits and squirrels are also mounted as hanging dead

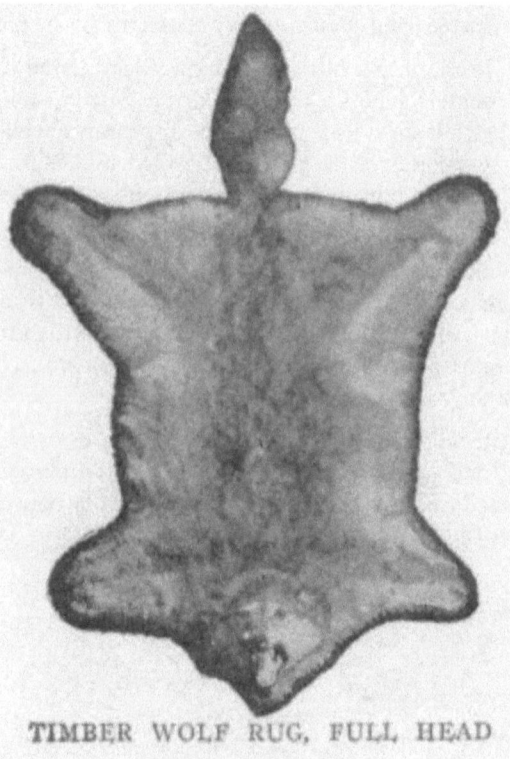

TIMBER WOLF RUG, FULL HEAD

game either in combination with some of the small game birds or separately. In selecting panels for this class of work use those finished in a contrasting color to the general tone of the specimens, a dark bird on a lighter panel and the reverse. On all panels and shields smooth rounded, beveled or Ogee edges are advisable. Small headings and intricate moulding are dust catchers. Wild cats, 'coons, foxes, coyotes, even bears and pumas gathered by night hunters and dog enthusiasts are usually best made up as more or less elaborate rugs. As wall and couch or chair hangings these have no trimming and often no lining except under the head. If for any reason the skins are unfit for this the heads can generally be used as wall mounts.

Room may be found for a few of the smaller specimens mounted whole but in the average home they are the bugbear of the housekeeper, early exiled to the attic. A friend of mine has his collection of small game birds, occupying the plate rail of his dining room, well out of the way and admired by many. Well mounted heads and antlers are suitable almost anywhere that they do not seem crowded. The famous East Room of the White House has some handsome examples. To make them answer a useful purpose they are made into hall racks, alone and in combination with feet. The makers of mounts offer a number of very attractive designs in the well-finished hard woods, some provided with plate glass mirrors. Fish make beautiful trophies which lend themselves particularly to wall decoration on panels or as framed medallions. How often the mounted trophy would save the fisherman's reputation for veracity. Perhaps their rapidly perishable nature accounts for the rarity of fish trophies. In conclusion I would say if you are a sportsman try the preliminary or entire preservation of some of your trophies, at least get them to the taxidermist in as good order as you can. Remember no matter how fine a specimen may have been, if allowed to be mutilated, become putrid or damaged, it can never be entirely repaired.

DEER HEAD HALL RACK.

The taxidermist must recall that exigencies of the field are responsible for neglect of many details and a nature loving sportsman is a friend worth having, who will share the contents of a seldom over-full purse with you in return for your best efforts.

CHAPTER XXV.

ODDS AND ENDS, TAXIDERMIC NOVELTIES.

There is almost no end to the useful and interesting things an ingenious person can turn out in this line. There is quite a demand for the preservation of the plumage of game birds for millinery use since the killing of other birds for this purpose was forbidden. Wings, tails, heads and breasts, principally, of grouse, pheasants and water fowl so used do not call up visions of starving nestlings. They need only to be skinned and poisoned as usual and pinned out to dry in the desired shape often loosely filling in and some cases wiring with rather small soft wire. When dry all raw edges or surfaces should be covered with pieces of cambric or lining canvas glued on.

Antlers and horns are sometimes worked up into armchairs and two pairs of small deer antlers turned upside down and screwed to a square of board make the foundation of a nice stool. Hat, gun and rod racks of feet, antlers and heads in various combinations are mentioned elsewhere and occasionally some one attempts an electrolier of antlers, mounted either on the heads or separately.

To do this grooves are chiseled out of the back of the antlers to receive insulated wire running to each point which is equipped with a light bulb. After placing the wires and bulbs and testing, the grooves are filled with "mache" or putty colored to match the other surface.

Peacock feather and fox tail dusters are fitted with buck horn handles or those made of fox or wild cat paws. Riding whips will look well with the same style handles.

Screens from mounted birds are highly ornamental, especially those of framed plush or satin on which birds of contrasting plumage are mounted in medallion style. It would be hard to find a more beautiful object than a snow white heron medallion on a black velvet screen framed in gold. These medallions are mounted by flattening the subject considerably so it is in little more than half relief.

A number of small birds may be mounted on a satin covered screen with embroidered branches and foliage. Some of the smaller fur bearers have been used in this way with success.

Some artists have specialized in grotesque mounting of small specimens, singly and in groups. Frogs, toads and squirrels are best suited to such caricature work.

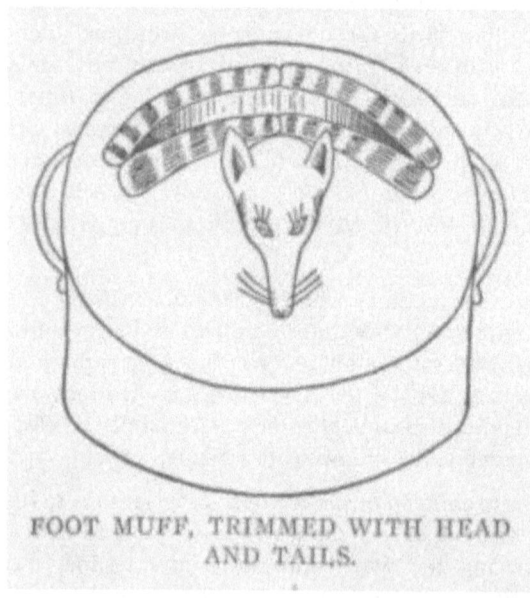

FOOT MUFF, TRIMMED WITH HEAD AND TAILS.

A foot muff can be made up from scraps of fur and will be appreciated in cold climates on long rides and indoors as well sometimes. To make this a covering of the size and shape of a foot stool is made of carpet or similar material. The bottom and sides are of this and the top of some short fur. A slit is made in this top and a bag of long fur or wool is sewed into the slit so when the muff or stool is loosely filled with tow and excelsior the feet may be thrust down into the fur lined pocket. The head of a fox or wildcat in half relief put on top, over the feet will give a finish to it.

A novelty in fur rugs is to mount the skin of some small animal in the center of a larger one of contrasting color. The so-called Plates of black goat are often so used with a fox, coon, or lynx in the center.

To do this mount the fox as for a half head rug, when dry and shaped cut out a paper pattern the exact size of it. Apply this pattern to the back of the goat plate, mark around it and cut out, leaving the opening a little smaller than pattern. Be sure pattern was right side up. Sew the skin in from the back, wad and line it. A felt trimming is unnecessary on this rug.

Match safes, candle holders, and similar things are made from the heads of fish and ducks with metal containers fastened in their open mouths. Monkeys, bear cubs and alligators mounted erect with card trays are quite striking while foxes or raccoons peering over the edge of umbrella jars or waste baskets are equally so. Many animals are mounted in Germany for advertising purposes, being either sold outright or rented by the month. Some of these are really a form of slot machine with coin actuated mechanisms while others are motor driven, attracting attention as moving displays always do. Bears and foxes on swings and seesaws and various small animals on merry-go-rounds are always attractive.

MONKEY CARD RECEIVER.

CHAPTER XXVI.

GROUPS AND GROUPING.

This subject is more of interest to the museum preparator than the home taxidermist, but a short consideration of it is not out of place here.

Many instructive and pleasing little groups of our smaller mammals and birds can be prepared for display in the home. Such groups usually require casing for protection but are well worth the trouble and expense.

Always try to make a group mean something. Let the subjects be feeding, fighting or occupied in any natural way. Family groups showing the male and female, adults and young, in the home surroundings are always good.

The seasonal groups of Spring, Autumn, Summer and Winter have been produced by most bird taxidermists at some time. Appropriate varieties of small birds are the blue birds for Spring; gold finches, Autumn; yellow birds or tanagers, Summer; snow birds, Winter. Framed with painted backgrounds and suitable accessories their shallow wall cases may be hung like pictures.

SQUIRRELS—GREY, RED, FLYING, GROUND (CHIP MUNK)

Never make the mistake of grouping animals that would never meet in natural circumstances or furnish them with incongruous surroundings.

The arrangement of groups for the exhibition cases of museums is very exacting as they are made open to the view on all sides. In order to judge of the affect such groups are modelled in miniature clay figures which are changed and re-arranged until satisfactory before the mounting is begun.

Such work is rather out of our province but an intelligent arrangement of two or more figures can be made to convey many more ideas than a single one would suggest.

Some of the most striking groups are those of the larger carnivora in combat, but they hardly possess the real value of painstaking life studies of some of our more familiar kindred of the wild.

CHAPTER XXVII.

ANIMAL ANATOMY.

A knowledge of this subject coupled with the necessary mechanical ability will enable their possessor to take place in the front ranks of taxidermists. Even if we have but little opportunity to study the anatomy of some of the rarer varieties of animal forms we can inform ourselves of certain typical features possessed in common by other more common members of the same great family or species.

Press and camera supplies us with much reliable information on the subject. Books on natural history, travels and sports were never so complete, interesting, and withal, so easy of access as they are nowadays.

A great help to the naturalist is a collection of pictures such as appear from time to time in periodicals. Back numbers of magazines on outdoor life and sports will contribute quantities of these, most of them reproduced from photographs and in a short time a large collection of such can be made. Packing these in the pockets of a letter file will keep them together, and at the same time make it possible to withdraw any one or more for inspection when wanted.

Photos of dead animals are not particularly valuable but casts always are; make them whenever opportunity offers. Not so much casts of the entire specimen as casts of various details.

Get a set of moulds of the noses of say deer, moose, domestic cattle and sheep and keep the resulting casts for reference. Their value will be apparent when mounting heads. Any sketches, however rough, will also be of use.

The circus and zoo will furnish feast days for the student of animal anatomy and pencil and camera may be used freely at both with the assurance of the best of treatment from officials and keepers.

A visit to the meat market will afford opportunity for study of the muscular system of the domestic animals.

The sculptor builds up his clay model unhampered by fur, feathers or bones and chisels out his statuary on a scale determined by himself while the taxidermist must not only construct his figures or manikins in correct proportions, but make them fit a certain skin. Hence it behooves him even more than the sculptor to be well grounded in at least the main principles of the anatomy of animals.

Birds in particular are a fruitful source of study, muffled as they are in feathers, when stripped presenting a very different appearance. To illustrate the value of a knowledge of avian anatomy I will mention an incident occurring many years ago at a large taxidermy establishment.

WATER FOWL HEAD.

Two of the frugal minded workmen having skinned a large plump duck laid the body minus head, feet, and wings aside to furnish a dinner next day. The porter regarding same as his perquisite abstracted and hid it. The first owners discovering it substituted the body of a large horned owl then in the process of mounting and so made all concerned happy. The porter bragging loudly next day of the fine duck he had done them out of, they were able to convince him of the truth only by exhibiting the duck remains as a part of their lunch.

CHAPTER XXVIII.

CASTING AND MODELLING.

One of the leading authorities in this country has aptly said, "The ideal taxidermist must be a combination of modeller and anatomist, naturalist, carpenter, blacksmith and painter. He must have the eye of an artist and the back of a hod carrier." This should not dismay the beginner for such casting and modelling as will be indispensable are comparatively simple.

In order to cast we must have molds and in our work these are chiefly of plaster. They are divided into two classes known as piece and waste molds. As the names indicate the latter is wasted or destroyed after making one copy while the piece variety can be used for a number of reproductions. The piece mold is divided into sections in such a way as will allow its removal without injury to either mold or cast. The waste mold is made from soft or fleshy objects which can be drawn from it in spite of projections known as undercuts.

As an illustration let us procure a cast of a deer's nose for reference in mounting the head later. For our purpose we wish a cast of the nose and lips, so with the head in the flesh at hand, the hair as far back as the corner of the mouth is coated with clay water to prevent the liquid plaster from penetrating and adhering. This done the head is propped up on the table and a rough box arranged around it, which will reach nearly to the mouth as the head is placed with nose uppermost. Pour sand in this box until only as much of the nose projects as is desired to cast.

Now mix in a bowl or basin a sufficient amount of water and plaster of paris to cover the surface of the deer's nose about inch thick. This should be of the consistency of cream and enough bluing or lampblack should be added to give it a decided tint.

If the skin of the lips and nose is disposed naturally the plaster may be ladled on with the spoon, endeavoring to get about an even coating. Wash any remaining bits from the dish and mix say twice the amount of plaster without coloring. Distribute this over the other and allow to harden, which they will do in about 20 minutes.

A little careful work will withdraw this mold from the nose and it may either be laid aside or used at once in making the cast.

To do this brush the inside with clay water and pour it full of plaster. Shake well to prevent bubbles and when hardened chisel away the mold. In doing this lay it on the lap or a cushion and chip off the mold. When the first layer (the colored one) appears work with caution to avoid marring the cast.

If a wire loop was inserted before the plaster hardened the cast may be hung on the wall for future reference.

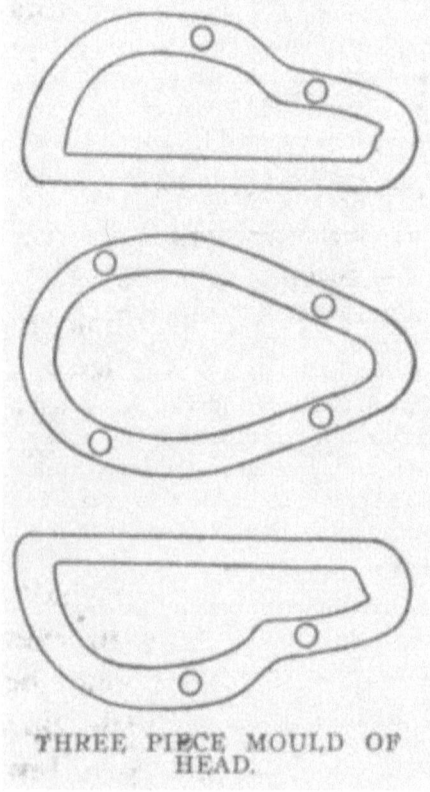
THREE PIECE MOULD OF HEAD.

The preparation of a piece mold is somewhat different. A mold can be made in two pieces of a round object like a ball and if each piece is exactly one-half, it will draw, because there is no point under which the plaster will hold. Any hollows or projections will form "under cuts" necessitating making the mold in a number of pieces that it may relieve or be lifted off the cast. Molds of heads from which to cast paper forms are often wanted and are easily made. With the skinned head of a fox, let us say, on the table, the lower part is embedded in fine sand or clay about on a line with the mouth. Cover half of the exposed upper part of the head also with clay. Pour to the depth of at least inch on the remainder.

Remove the clay from the other half of the face, and after countersinking two or three shallow holes in the edge of that part of mold already made and coating that edge with clay water, pour plaster for the second piece of mold. When this hardens pick up head from its bedding of sand or clay and turn over so the final piece of model can be made.

Always coat the edges with shellac or clay water to prevent adhesion and countersink a few holes for dowels to aid in holding the pieces in place. Dry out thoroughly and shellac the whole interior and joining edges. If it is slightly oiled before using a great number of casts may be made from it. This will give us a complete cast of a fox head with closed mouth.

A shorter method to obtain molds of the upper part of the head and face for making paper half-head forms, is to imbed in sand or clay as directed and stick a piece of stout thread or cord along the central lines of the head and face. A little clay will hold this in place and there should be a few inches surplus at each end. Mix the plaster and cover the entire top and sides of the head with it. Just as the plaster begins to harden draw the thread upward through the stiffening plaster cutting it in two parts which are easily removed when hard. When dry coat with shellac, tie together and they are ready for use.

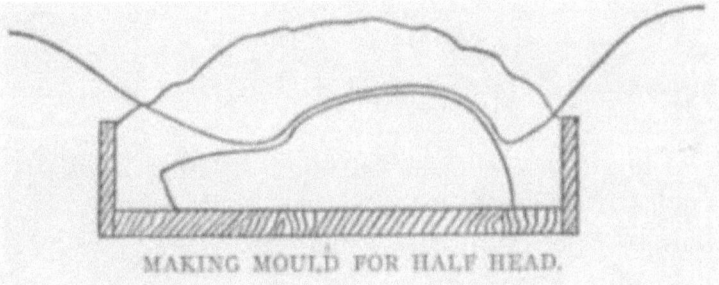

MAKING MOULD FOR HALF HEAD.

To cast half head forms soak some paper and after coating one side with paste, press into the mold with the fingers. The first layer should be quite soft so as to crowd into all depressions. About six layers of building paper is thick enough for a fox head size. When dry cut the cords and detach the mold.

Molds for deer head forms are made in two pieces, one for each side of the head, and are necessarily not carried completely around the antlers. This gap is just filled in the head form by the plate of bone bearing the antlers, which is sawed from the skull.

The entire neck may be molded in connection with the head if desired. Gelatine and compositions of glue and wax are used for molds where fine definition is desirable, and wax as well as plaster and paper for making casts. The ground up paper pulp is used for many casts, pressing it into mold with fingers and spatulas.

Clay is the stand-by of the taxidermist modeller. That furnished by art dealers is best, but for common use potter's clay is all that is necessary. A little glue mixed in plaster delays its setting and makes it harder when dry. Good papier mache is one of the best materials for much modelling and wax for very fine work. Tools for this work may be purchased or home made of wood, bone or metal.

Many forms of fishes and reptiles are difficult or impossible to mount by ordinary methods. On these the caster and modeller may work his will, and if he also possesses a good eye for color the results may be of the best. As an indisputable record of anatomy even a poor cast is valuable.

CHAPTER XXIX.

MARKET TROPHY HUNTING.

In this country and day of conservation this would seem like a delicate subject to attack. The hunter for the trophy market a few years back was slaying elk, mountain sheep, moose, deer, or antelope indiscriminately.

DEER FOOT INK WELL AND PEN RACK.

While modern game laws have changed or at least modified this I can see no reason why a hunter who is entitled to a certain head of game per season should not utilize them fully by preparation and sale to others who have not similar opportunities.

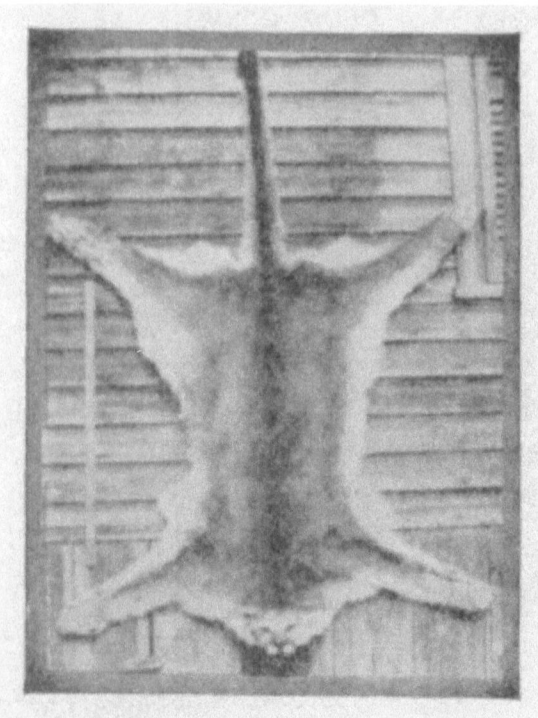

MOUNTAIN LION OR PUMA HIDE.

What would often be left in the woods as useless, as indeed it would be for food purposes, is transformed into a beautiful and decorative article of considerable commercial value. Often things being equal the trophy hunter will avoid killing young and female game animals on account of the worthlessness of their heads as trophy if not for any ethical reason.

While the day of trophy hunting as a business in the United States is past probably, by preserving such heads, horns, feet and skins as come in his way the trapper, prospector and settler can often add considerably to his income. For instance, from one to five deer may be legally killed in different states. If two good heads are taken, worth say $15.00 and $20.00 each when prepared, that sum would go far towards paying the expenses of an enjoyable outing.

The fur trapper will frequently take some animal the skin of which may for many reasons be of little value. The puma or mountain lion is such a one, worth but $2.00 or $3.00 usually, the mounted head is a striking wall ornament and the skin is suitable for couch or floor.

Though fur dealers will make some deduction from the regular prices on skins from which the heads are removed, it is vastly more profitable to retain them and preserve as trophies.

Horns and antlers and head skins or scalps of all our large game have a certain value either separately or together. Mounted heads, damaged by moth create a demand for extra scalps and separate antlers are often called for. Extra large heads or antlers of freakish formation seem to possess a special fascination for the public.

SPRING LAMB? COON HEAD.

Commercial fishermen handling fish in large numbers would do well to preserve at least a few of the more notable specimens of their catch.

In some localities there is every summer an opportunity to supply "rusticators" with rattlesnake skins which may be prepared for wall decorations or use as belts, hat bands, card-cases, and neck ties. They should be packed in salt until tanned as drying out while in the raw state is apt to spoil them. On account of the snake's habit of shedding its skin at varying intervals, dressing snake skins is rather of the nature of a lottery. The dressed skins should be made up with a backing of some other leather as it is apt to possess but little strength of itself.

In localities where the tarpon, tuna, muscallonge, and other large fish are caught it is well to keep some good specimens on hand as such are often in demand to substantiate a fish story.

In a word, gather and preserve some of the best objects of animated nature your locality affords, whether fur, fin or feather.

CHAPTER XXX.

COLLECTING AND MOUNTING FOR SALE.

Commercial taxidermy is roughly divided in two branches, custom work, and collecting and mounting for sale. For the first you need some fixed place of business easy of access to the public and convenient to lines of transportation. The latter may be taken up anywhere if a demand has been noted and a market assured or in prospect.

Travelers in little known parts of the country often pay their expenses or even gain considerable profit by collecting desirable specimens of animal life. As a side line on pleasure trips it is sometimes remunerative. Woodsmen and fishermen will often find it to pay better to preserve for mounting part of their game at least.

The sales end of the proposition is the most difficult for the outdoor man. Such work has not the fixed (?) value of furs and meat. There are a number of dealers in naturalists' material who aim to keep on hand a pretty complete stock of specimens for museum purposes. Correspondence with these will procure their want lists.

BOOKCASE ORNAMENTS—CROW; ALLIGATOR (Photograph from the Everglades); OWL.

Many more deal in unmounted trophies of heads, horns and rug skins. Occasionally an order for small and common species may be secured from some school or college. Such institutions will often place an order for desirable material with a prospective traveler.

Finally it is well to mount a good specimen or two of almost any variety on general principles. It is astonishing how difficult it is to procure some very common species on the spur of the moment. If you accumulate a number of nicely done and attractive specimens it is possible to secure their sale on commission.

As such things are apt to draw attention as a window or wall display some druggist, sporting goods dealer or other business man may be glad to aid in their disposal. In or near a game country the local hotels will help advertise you by giving wall space in dining room or office to suitable pieces accompanied by a business card. Donations to libraries, schools and other public and semi-public institutions will keep you more or less in the public mind.

Endeavor to fill any orders you receive even if obliged to purchase at such rates that no profit remains.

Do not diminish the animal life of your locality by collecting everything you can lay your hands on. It would be time misspent and mostly unrewarded.

CHAPTER XXXI.

PRICES FOR WORK.

To those who hope to coin spare hours into dollars and cents, or others who must make a hobby pay its own expenses at least, an important question is, what is my work worth?

And one will concede that a taxidermist should receive at least as much as a skilled mechanic and the experts both in commercial and museum work are sometimes (not always) highly paid.

What seems the fairest method of compensation is by "piece work" and most custom taxidermy is handled on that basis. Most professionals have a regular scale of prices which, while necessarily more or less elastic, will give the public an estimate of cost.

The schedule which I give is, I think, about that in use in the Eastern States. The outside prices are for extra large specimens or those mishandled or injured so as to require an extra expenditure of time to give satisfaction.

PRICES FOR MOUNTING SPECIMENS.

BIRDS.

Small Wrens, Canary,	$ 1.00 to	$ 1.50
Robins, Jays, and similar,	2.00 to	2.50
Medium Quail, Snipe, Dove, Woodcock,	2.50 to	3.00
Large Crow, Grouse, Duck,	3.00 to	4.00
Larger Horned Owls, Fish Hawk, etc.,	4.50 to	5.00
Eagle, Turkey, Pea Fowl,	10.00 to	15.00

Birds with spread wings add 25 per cent.

MAKING BIRD SKINS.

Small up to size of Sparrow,	.20
Robin, Jay, etc.,	.25
Pigeon, Hawk, and similar,	.35
Screech Owl, Green Heron,	.50
Crow, Teal,	.75
Large Hawks, Ducks,	$ 1.00
Herring Gull, Eider Duck,	1.25

Great Horned Owl, Fish Hawk,		1.50
Eagle, Goose, Swan,		2.50

WHOLE ANIMALS.

Mouse, Mole, Chipmunk,	$ 1.50 to	$ 2.00
Squirrels, Weasels,	2.00 to	3.00
Mink, Muskrat, Opossum, Rabbit,	3.00 to	4.00
Skunk, Woodchuck,	4.00 to	5.00
Coon, Fox, Wildcat,	6.00 to	10.00
Dogs,	10.00 to	35.00
Domestic Sheep,	25.00 to	40.00
Bear, Mountain Lion,	20.00 to	75.00
Deer, Antelope,	30.00 to	75.00

Price on whole mounted specimens include rustic stands, stumps, or rock work.

HEADS.

Elk, Moose, Steer,	$ 20.00 to	40.00
Caribou, Mountain Sheep,	15.00 to	25.00
Deer (buck), Antelope,	7.50 to	12.00
Deer (small), common sheep,	5.00 to	10.00
Bears,	7.50 to	15.00
Wolf,	5.00 to	7.50
Fox, Wildcat, Raccoon, etc.,	4.00 to	6.00
Hawks, Owls, Eagles,	2.00 to	3.00
Fish,	2.00 to	5.00

Suitable shields or panels are included.

FISH, REPTILES, ETC.

Small fish,	$ 2.00 to	5.00
Medium, Bass, etc.,	5.00 to	10.00
Large, Tarpon, Salmon,	10.00 to	25.00
Snakes, as to size,	5.00 to	25.00
Alligators,	1.50 to	25.00

MOUNTING HORNS, INCLUDING SHIELDS.

Deer,	$ 2.50 to	5.00
African Horns,	2.50 to	10.00
Cow, Steer,	2.50 to	5.00

Caribou,	3.50 to	7.00
Moose, Elk,	5.00 to	10.00

SKINS.

First column shows cost of tanning only; second of tanning, mounting head and lining as rug; third of complete rug with open mouth.

	$	$	$
Black Bear,	4.00	10.00	15.00
Mountain Lion, Jaguar,	3.00	10.00	15.00
Tiger,	5.00	15.00	20.00
Wolf,	2.00	8.00	12.00
Coyote, Lynx,	1.50	7.50	10.00
Fox, Wild Cat, Coon, House Cat	1.00	5.00	6.00
Sheep,	1.50	....	....
Goat,	1.50	8.00	....
Deer,	2.50	10.00	....
Opossum, Muskrat,	.50	....	....
Mink,	.75	....	....
Snake,	1.00 and up		
Alligator,	2.00 and up		

NOVELTIES.

Deer Feet, each,	$ 2.00	to	3.00
Moose and Elk Feet,	3.00	to	4.00
Including fittings.			

ROBES.

According to size and variety of skins from $15.00 to $25.00 including tanning, sewing up and linings. The smaller skins of course require the most work.

Domesticated animals, dogs, cats, cage birds, etc., are mounted at the rates for similar sized wild specimens. Inasmuch as they are of value only for associations most taxidermists require a small advance payment on pet animals before commencing work; other work is usually C. O. D.

A discount of 10 to 20 per cent is often made for large quantity or to those in the fur trade who may be so induced to secure orders.

It would pay for at least one person in every furriers shop to have a knowledge of taxidermy and a connection with some dealer in sportsmen's goods is often of advantage.

Much of this matter of prices must be left to your own judgment. Often a fair profit can be made on work taken at a low figure during the "off season." Perishable work demanding instant attention should receive the best pay and pieces which may be picked up in odd moments, thus using time otherwise valueless, may be figured near the foot of the scale. The public appreciates work thoroughly done and it is the very best advertisement.

www.ingramcontent.com/pod-product-compliance
Lightning Source LLC
Chambersburg PA
CBHW031624210526
45464CB00004B/1732